2-13-68

Science and the Mass Media

SCIENCE AND

THE MASS MEDIA

by *Hillier Krieghbaum*

PROFESSOR OF JOURNALISM
WASHINGTON SQUARE COLLEGE OF ARTS AND SCIENCE
NEW YORK UNIVERSITY

NEW YORK UNIVERSITY PRESS *New York*

UNIVERSITY OF LONDON PRESS LIMITED *London*

1967

1444372

PREFACE

This book was written for those with a concern or interest in the public understanding of science, technology, and medicine. It is hoped that the audience will include those scientists, engineers, and physicians who are not satisfied to work at research and development without thinking of the consequences of their activities; those practitioners, teachers, and students of journalism who want to know more about this phase of contemporary reporting; and those considering a career in science journalism for themselves or for those they advise.

The author wishes to thank the many science writers who contributed material in this volume and especially the National Association of Science Writers, Inc., whose *Newsletter* and *Clipsheet* provided so many quotations. Appreciation also is expressed for permission to quote from publications of

American Academy of Arts and Sciences; American Association for the Advancement of Science; Association for Education in Journalism; the Council for the Advancement of Science Writing, Inc.; and Sigma Delta Chi, professional journalism society, and also from *Columbia Journalism Review* and *Television Quarterly*.

Hillier Krieghbaum

May 3, 1967
Mamaroneck, New York

CONTENTS

ix

Science and the Mass Media

1

WHY BOTHER?

If an outsider were to listen briefly to a group of scientists, technicians, and reporters discussing the difficulties of telling the public about what was going on, he undoubtedly would comment: "Why bother? The results hardly seem worth the effort."

While there may be many reasons for reaching that conclusion on short notice (and with insufficient background), it would really reflect the outsider's naïveté rather than any informed evaluation. Although it may seem redundant to a sophisticated observer, much may be achieved by briefly going over the arguments for informing the public about science, technology, and medicine.

But since there are just about as many ways to report these developments as there are ingenious writers to do so, we probably should start with a definition of the word "science." For this book, all the various interpretations attached to the word are lumped together — unless specifically delineated in another way. Thus my usage will include so-called "pure" science or basic research, applied science or development, technology, engineering, medicine, and public health. It is a bit unfortunate that the word "science" is used in an all-inclusive sense as well as in an exclusive sense to mean basic research, the so-called "pure" science which former Secretary of Defense Charles E. Wilson described condescendingly as what you do "when you don't know what you are doing."

As a rationale for choosing the embracing definition, if such is needed, I can quote Dr. Warren Weaver's comment in a paper for the President's Commission on National Goals in 1960. The mathematician and former vice-president of both the Rockefeller and Alfred P. Sloan Foundations wrote that the distinction between "pure" and "applied" research was interpreted differently by scientists themselves and "is often more a matter of temperament and motivation than it is of procedure or result." He argued that both are "of the highest importance and it is silly to view one as more dignified and worthy than the other."

One other definition also may be useful: Science writing, for purposes of this book, refers to the dissemination of such news and information to the general public through the mass media. It does not include how scientists, technicians, engineers, or physicians communicate with their colleagues and peers — unless, of course, this is done through daily newspapers, news magazines, radio or television programs aimed at the average individual, not the specialist.

Now let us examine the traditional arguments for informing the general public — the curious layman — about developments in science.

First, the applications of science now underpin our contemporary way of life. This has been exemplified in our high-speed air transportation, our electronic communications, our public health activities, our general economic well-being, and our national defense. It is to these practical applications that the public turns when it thinks of the benefits of what some have called our "scientific culture."

This predisposition was documented in a 1957 survey of the public's attitudes toward science and scientists made by the National Association of Science Writers, a professional group of more than 800 practitioners. When a national sample of 1,919 respondents was asked, "All things considered, would you say that the world is better off or worse off because of science?", 83 per cent unhesitatingly replied that it was better off and 5 per cent gave qualified approval. Only 2 per cent said the world was unquestionably worse off and 1 per cent reported qualified disapproval. When a probing question inquired for the reasons behind their opinions, almost half of the group (49 per cent) cited improved health and better medical treatment; 45 per cent mentioned better living standards; and 19 per cent talked of industrial and technological improvements.

In other words, the public at large thinks first of the practical aspects — longer life, better living, and gadgets — rather than of the more philosophical and cultural gains often discussed by scientists.

When they are talking to others than their colleagues, even scientists fall back on these arguments. To illustrate, when 15 "distinguished" scientists, as the president of the National Academy of Sciences described them, were asked in 1965 to tell the House Committee on Science and Aeronautics their ideas on how much money should be spent on federal support of science and how these funds should be split, the summary reported that "the argument for public support of basic science because it is a distinctive element of our culture is conceded by most of the panelists to be less persuasive than is the argument based on useful application of basic science."

Science has become an absolutely essential ingredient for a

national defense program, as Sir Winston Churchill made clear
in discussing World War II:

> Unless British science had proved superior to German
> and unless its strange sinister resources had been effec-
> tively brought to bear in the struggle for survival, we
> might well have been defeated, and being defeated, de-
> stroyed.

During more recent years, this argument has received further
reinforcement with the development of intercontinental ballis-
tic missiles and plans for manned space stations. What has been
true for the past will undoubtedly prove true for the future.

Despite popular enchantment with the applications of science, many of the more important and complicated problems
facing United States citizens today are heavily intertwined
with science and technology. They cannot be approached
soundly without an appreciation of their scientific implications.
Often these require understanding subtle nuances grounded in
basic research. To illustrate, think of the background needed to
discuss intelligently such topics as population explosion and
birth control, uses and abuses of automation, pollution of the
natural environment, water conservation and irrigation, uses
of insecticides and pesticides, testing of nuclear weapons,
peacetime uses of atomic power, and the relationship of cigarette smoking and cancer. Before the final decision is made in
any democracy, the people should have an opportunity to consider the problems; at least the public has to pick political
leaders to find the answers, generally after campaign debates.

In 1959, the President's Science Advisory Committee in *Education for the Age of Science* pointed out that such social and
political problems as those just cited and the others that will
arise in the future, are too urgent to await the enlightenment
of the electorate that will slowly come about with improvements in the nation's educational system. That report concluded:

There is, therefore, no escape from the urgency of providing highgrade and plentiful adult education in science now, planned for those who are unprepared even in the fundamentals.

Yet the "adult education" available to a majority of the public is that provided by the mass media. This fact remains true despite the needed and planned expansion in conventional adult education courses. Thus, if the public is to make wise and intelligent choices, it needs to know its science now and the most accessible way for it to get this is from printed media, radio, television, and film.

Despite some peevish complaints and anguished attacks from those who feel the humanities have been submerged and neglected, the generally approving acceptance of science and its applications as vital elements of contemporary living has become, since World War II, almost stereotyped. Here are some illustrative comments from both scientists and laymen.

Dr. Alvin M. Weinberg, director of the Oak Ridge (Tenn.) National Laboratory, said in a 1961 speech:

> When history looks at the 20th century, she will see science and technology as its theme; she will find in the monuments of Big Science — the huge rockets, the high-energy accelerators, the high-flux research reactors — symbols of our time just as surely as Notre Dame is a symbol of the Middle Ages.

In his 1959 Rede Lecture, which became the basis for the widely quoted book, *The Two Cultures and the Scientific Revolution*, C. P. Snow (now Lord Snow) indulged in what he called "a piece of shorthand" and claimed that today's scientists "had the future in their bones." The British author and scientist also quoted Sir Ernest Rutherford as trumpeting, "This is the heroic age of science! This is the Elizabethan age!" Then he added his own evaluation:

What is hard for the literary intellectual to understand, imaginatively or intellectually, is that he was absolutely right.

In December, 1963, the House Subcommittee on Science, Research, and Development could generalize:

Today, no great perception is required to understand the exploding importance of science and technology to the welfare of America, present and future.

And the chairman of the subcommittee, Rep. Emilio Q. Daddario, Dem., Conn., wrote for the same publication:

It is already apparent that we as a Nation must maintain not only present but future excellence in science. The Nation's well-being and security depend on it.

The public and the Congress must make a serious and determined effort to inform themselves of the possibilities — and the risks — of science. This requires an understanding acquired through attention and diligence, of the basic principles, as well as awareness of the latest advances.

The same House Subcommittee on Science, Research, and Development, in March, 1965, expounded:

Any nation wishing to survive in this modern world, much less hold a position of importance, must equip its citizenry with the knowledge and skills demanded by this scientific and technological age and with the understanding to cope with future technological developments. The position of the United States, as the leader of the free world, is dependent upon how well we are able to provide our people with the intellectual tools demanded by our increasingly complex technological society.

Dr. Herbert J. Muller, Distinguished Service Professor at Indiana University, wrote in the concluding volume of his History of Freedom series, *Freedom in the Modern World* (1966), that science may always look like the villain in the modern intellectual drama because of its "primary disruptive force" but that it remains "the basic source of our extraordinary intellectual resources." Then he added:

> Like it or not, science is an irreversible force, at least as long as civilization endures; for no educated man can ever think as if it had not been, and there is no chance whatever of calling a halt to it, any more than of junking all the knowledge it has given up about the universe and man. The sensible policy would seem to be an effort to feel more at home with it, immediately for the sake of historical understanding, ultimately for the sake of maintaining our freedom.

David Sarnoff, board chairman of the Radio Corporation of America who rode one of the most popular waves of scientific application, said in a 1963 speech:

> The world of your lifetime and mine is being changed politically, economically, socially, technically, even geographically, by epoch-making scientific breakthroughs. There is not a man alive today who is unaffected by the new scientific discoveries and their applications. . . .
> In recent decades, the products of the research laboratory and drawing board have poured forth at a bewildering pace. There has in fact been more progress in discovery and development of scientific knowledge and in modification of our environment since the beginning of the present century than in all the prior millennia of recorded history. . . .
> What we have done in the past half century is to delineate the major areas of scientific conquest and to sketch

their potentialities. Their fullest development is still to come.

The New York Times' James Reston, one of the most respected commentators in journalism today, reported that a journey through European capitals late in 1964 left him with the impression that "it is not the politicians but the new scientists, engineers and economists who are shaping the modern world." Then he continued:

> Everywhere the politicians and even the statesmen are merely scrambling to deal with the revolutions in weapons, agriculture and industry created by the scientists and the engineers. The latter have transformed man's capacity to give life, to sustain and prolong life, and to take life; and the politicians no longer find that they can deal with all the complexities and ambiguities in ideological terms.

Not only good health but life itself may depend on "the communication of lifesaving knowledge to the potential victim or to the physician who treats him." As the President's Commission on Health Disease, Cancer, and Stroke commented in February, 1965, "It has been said that knowledge is power. In heath, it is the power of life and death."

With just a slight touch of professional chauvinism, the National Association of Science Writers distributed *A Guide to Careers in Science Writing* in the middle 1960's with a paragraph which read:

> Nobody really doubts any longer the importance of science in our lives and to the future of the world. The last skeptics surrendered after Sputnik. Today people who never got past freshman biology have heard of deoxyribonucleic acid — even if they aren't precisely sure what it means.

Second, the United States taxpayer is the main supporter of scientific research and development today. The expansion in the federal government's contribution to such funds has been truly astronomical. In the 25 years from Fiscal Year 1940 or just before Pearl Harbor to Fiscal Year 1965, appropriations rose from $74,000,000 to $15,287,000,000. This increase is more than two hundredfold in a single generation. Looked at an- other way — in terms of the total federal expenditures — the research and development funds shot up from eight mills to 15.6 cents out of every tax dollar spent.

By five-year intervals, figures compiled by the National Sci- ense Foundation show:

FISCAL YEAR	TOTAL FEDERAL EXPENDITURES	TOTAL RESEARCH AND DEVELOP- MENT FUNDS	R AND D PERCENTAGE OF TOTAL SPENDING
1940	$ 9,055,000,000	$ 74,000,000	0.8
1945	98,303,000,000	1,591,000,000	1.6
1950	39,544,000,000	1,083,000,000	2.7
1955	64,389,000,000	3,308,000,000	5.1
1960	76,539,000,000	7,738,000,000	10.1
1965	97,900,000,000	15,287,000,000	15.6

If the piper's tune is really called by those who pay him (and this is probably more true in political affairs than in other fields), federal appropriations long ago attained the level where informed background was necessary for both govern- ment officials, especially members of Congress, and the lay public which elects them. Both officials and the public need to listen to briefings from experts but they should not be captives of these specialists. To become hostages of any group, however benevolent its intentions appear, is to expose oneself to a chance to be misled and exploited, sometimes with dangerous conse- quences.

Because a culture cannot overly departmentalize its components without danger, science should not be treated as something that requires only casual attention. As Dr. J. Bronowski argued in *Science and Human Values* (1956):

> There is no more threatening and no more degrading doctrine than the fancy that somehow we may shelve the responsibility for making the decisions of our society by passing it to a few scientists armored with a special magic. . . . The world today is made, it is powered by science; for any man to abdicate an interest in science is to walk with open eyes towards slavery.

Along the same lines, Dr. Polykarp Kusch, Columbia University professor and 1955 Nobel Laureate in physics, has warned of the dangers of "an abdication of the right and the responsibility of every man to participate in forming the fabric of his society." Thus, he told the 1961 Pulitzer Prize jurors, the press faces the crucial job of "having man become attuned to a world that is heavily conditioned by science and technology."

Such elevation of decision-making to the "experts" without broad social assessments may be bad for science itself. The Committee on Science in the Promotion of Human Welfare of the American Association for the Advancement of Science reported in 1964 that its survey showed the pressure of social demands and the rapid modern expansion of science had led to "serious erosions in the integrity of science." The report commented further:

> To arrogate to science that which belongs to the judgment of society or to the conscience of the individual inevitably weakens the integrity of science.

When a physicist is speaking about the best site for a multimillion dollar accelerator, for instance, the listener should know at least enough to separate those statements grounded on accepted scientific facts from those that are highly subjective, sectional, or institutional, pleadings. As "Big Science" with increasingly expensive installations comes to the fore, this need for discrimination and even sophistication mounts — just as does the amount of money involved.

But can scientists really justify for the laymen, who must pay most of the tax bills, expenditures of $500,000,000 annually for high-energy particle accelerators during each year of the 1970's, as was proposed by one group of physicists in 1963? Is this not a high price for what has been called "a purely intellectual expedition into the unknown," no matter how exciting and stimulating this quest into the nature of matter may appear a generation or two in the future? Even scientists in fields other than physics remain unconvinced of the efficacy of the expenditures.

In other words, does one not eventually have to set upper limits to appropriations for science and development? Weinberg claims that if both the gross national product rate and the rate of research and development spending continue their recent scale of increases, then the United States will be spending *all* its money on science and technology by A.D. 2030. Obviously such trends will not continue uncurbed and the real crunch will be to make valid decisions on how to split the money bag among the various science applicants. Here the public needs to wield its own considerable influence — after being adequately informed. One of the first points will be to distinguish explicitly between research and development.

When President John F. Kennedy proposed the lunar landing before the end of this decade, Congress sped through initial appropriations for Project Apollo with its estimated $40,000,-000,000 eventual price tag at a speed that startled even some congressmen. Only later did the public (and probably many congressmen) learn that a considerable number of scientists believed the money might better be spent here on earth — to further the search for cures of cancer and heart disease, to improve mass education, or even to feed some of the millions of undernourished in the world. Two months before President Kennedy's assassination, Columnist Walter Lippmann wrote of mistakes that had "transformed what is an immensely fascinating scientific experiment into a morbid and vulgar stunt." Certainly we can wonder if Congress would have made the same decision if there had been more vigorous debate by the experts in the first place and then more considered public reactions. But the lunar commitment had been made before all the world, and the nation moved to keep it — if possible — even after belated questioning of the initial decision and of the arguments that some experts had used to win Kennedy Administration support.

Unless there is real understanding of science and technology (not necessarily a comprehension of nuts and bolts details and specialization but rather a backdrop of basic knowledge and an intelligent appreciation of goals), our cherished forms of traditional decision-making in a democracy face new and dangerous threats. For example, some scientists are now discussing seriously a sort of guaranteed annual salary for competent workers who do not require the expensive machinery of "Big Science" so that they might be free to range wherever they think needed basic research should be done. This proposal stimulates the imagination of both scientists and intelligent laymen, but all scientists might not be entirely disinterested during testimony they gave on this idea.

Larger and larger sums of money will probably be earmarked for science, as indicated. Although the rate of increase has slowed and it may fluctuate, a sustained plateau or decline in science appropriations does not seem at hand for the immediate future. But when that time approaches, if it does, the scientific community might well contemplate what Dr. Caryl P. Haskins, president of the Carnegie Institution of Washington, said in his 1965–1966 report:

> When federal support of any American institution or activity reaches such levels, it is inevitable that the control of that activity or institution, the judgment of its directions, the shaping of its courses, will become matters of public, indeed of political, judgment and decision. Close and continuing public scrutiny of an enterprise to which

so large a portion of the national fortune is dedicated become inevitable — and indeed essential. . . .

Whenever demands from other sectors of the economy force a reduction in the federal funds for research and development, or even when the rate of increase of expenditures in these areas must be appreciably curtailed, there is serious risk that, in the absence of adequate public understanding, pressures to cut off support of fundamental work, however basic and important it may be in the long term, in favor of the retention of the immediately obvious, may mount to the point of irresistibility. And we are far from having achieved an adequate public understanding today.

Third, science is "an adventure of the human spirit," as Weaver described it. Many scientists themselves speak ecstatically and almost reverently of the beauties and excitement of science and basic research. One went so far as to call modern science "mankind's most intellectual attainment." And Weaver added, "All citizens would be given a richer inner life if they could have a chance to appreciate the true nature of science and the scientific attitude."

Dr. Donald F. Hornig, Princeton University professor who became special assistant for science and technology to President Lyndon B. Johnson, commented on this aspect of science before the thirtieth anniversary meeting of the National Association of Science Writers at Philadelphia in 1964 when he said:

What distinguishes science from other creative activity is that it has proved possible not only to invent ideas which bring beauty, elegance and order to the world of nature but to build them one on another. It has developed a mode of communication which involves many minds in a common endeavor, united in a common inspiration and seeking a kind of truth which, being testable, can withstand differences of ideology, race and nationality. . . .

I do not mean to give the impression that science has

solved all our problems or that it will ever be capable of it. What has emerged is that there is a set of intellectual frontiers which can be pushed back systematically, that ideas can be constructed in ways which constantly expand what has been done before, so that at the present time the horizons are further away than in the past and opening up further all the time. As our sight improves, we constantly recognize more problems still to be solved, and that is part of what makes it all so exciting.

Much earlier, Albert Einstein pointed out the other side of the argument for public involvement when he said:

It is of great importance that the general public be given an opportunity to experience, consciously and intelligently, the efforts and results of scientific research. It is not sufficient that each result be taken up, elaborated, and applied by a few specialists in the field. Restricting the body of knowledge to a small group deadens the philosophical spirit of a people and leads to spiritual poverty.

The humanistic scholar, Dr. Jacques Barzun of Columbia University, wrote in his provocative and far from apologistic critique, *Science: The Glorious Entertainment* (1964):

Out of man's mind in free play comes the creation Science. It renews itself, like the generations, thanks to an activity which is the best game of *homo ludens:* science is in the strictest and best sense a glorious entertainment. And I intend glorious to mean at once magnificent and deserving of glory.

Dr. J. Robert Oppenheimer, then director of the Institute for Advanced Study at Princeton, made the point at the 1963 centennial of the National Academy of Sciences when he advocated that young people and "older people who are young in heart" should have a chance to share the experiences of scientific discovery along with the scientists because it is "a good and beautiful experience, and an unforgettable one." He said it is not arrogance but simply human to wish these pleasures for as many of our fellows as can have them.

He also said:

> We have a modest part to play in history, and the barriers between us and the men of affairs, the statesmen, the artists, the lawyers, with whom we should be talking, could perhaps be markedly reduced if more of them knew a little of what we were up to, knew it with pleasure and some confidence; and if we were prepared to recognize both the important analogies between what moves us to act and to know, and the extraordinary and special quality of our experiences and our communication about it with one another. I have often thought that with the historic game so grand and so uncertain, we should not dismiss any help, even of that small part which we could play.

At the same centennial ceremonies of the National Academy of Sciences, President Kennedy touched on this aspect of science when he remarked:

> I can imagine no period in the long history of the world where it would be more exciting and more rewarding than in the field today of scientific exploration.

Is there evidence that laymen can share this enthusiasm of the scientists? Some feel so. For instance, during a 1963 address at the Rockefeller Institute in New York City, Dr. Frederick Seitz, president of the National Academy of Science, speculated about a new phase in the evolution of science and technology where a broader public interest would transcend the purely practical. He continued:

The new phase will be based on the combination of a number of motivations meaningful to the average man, such as pride, competition, sportsmanship, amusement, curiosity, as well as the purely practical. The public support of the space program clearly has some aspects of such nonpractical motivation along with the practical ones. One can conceive of many other programs of a similar type which could involve a total annual investment in research, development, testing, and evaluation of the order of a quarter or even half of the gross national product, that is, about ten times as much as at present.

And it is this area — which Gerald Piel, publisher of *Scientific American*, called "the illuminating and the beautiful ideas that come out of the work of science" — that some journalists believe should be the main arena for science popularization. Piel explained that the current preoccupation of science journalists with information should give way to "popularization of the objectives, the methods, and the spirit of science." Only when this takes place, he said, will the public support the advance of science for motives other than utility.

Improvement of the scientific background of laymen would spill over to clarify popular thinking about the methods of science too, as Dr. James B. Conant, former Harvard president, emphasized in his book, *On Understanding Science* (1947). Such an improvement could lay "the basis for a better discussion of the way in which rational methods may be applied to the study and solution of human problems." When and if this actually happens, it would not be an inconsiderable asset to a society that cherishes reason.

If Weinberg is correct about medieval cathedrals and the instrumentation of "Big Science" today, then possibly recognition of and involvement with science by laymen compares with the appreciation of these great churches by individuals who are neither architects nor contractors.

These intangible appeals might be difficult to measure by any physical scale or to ascertain by psychological analysis, but they nevertheless may provide a triggering motivation that in the long run would be more effective in moving the general public toward truer concepts of science than either its massive applications or the expansive federal appropriations.

2

THE MULTIFARIOUS NATURE
OF SCIENCE NEWS

The journalist's interest in reporting on science to the general reader is no Johnny-come-lately infatuation. It goes back to the initial (and, as it turned out, the only) issue of the first American newspaper. When Benjamin Harris, publisher of *Publick Occurrences*, which was dated September 25, 1690, assembled items to inform Massachusetts colonists of what was happening and, as he wrote, to quiet "many *False Reports*, maliciously made, and spread among us," he included two paragraphs that qualified as the first American newspaper reporting of science news.

In view of the colonists' concern about the then prevailing

"Fevers," it was not surprising that Harris printed such information — even though his career did not show him to be a particularly astute or creative newsman. Since they have historical value, let us look at these two paragraphs, which began on the first page and continued on to the second. They read:

> Epidemical *Fevers* and *Agues* grow very common, in some parts of the Country, whereof, tho' many dye not, yet they are sorely unfitted for their imployments; but in some parts a more *malignant Fever* seems to prevail in such sort that it usually goes thro' a Family where it comes, and proves mortal unto many.
>
> The *Small pox* which has been raging in *Boston,* after a manner very Extraordinary, is now very much abated. It is thought that far more have been sick of it than were visited of it, when it raged so much twelve years ago, nevertheless it has not been so Mortal. The number of them that have [Page 2] dyed in *Boston* by this last Visitation is about *three hundred and twenty,* which is not perhaps half so many as fell by the former. The time of its being most *General,* was in the Months *June, July* and *August,* then 'twas that sometimes in some one Congregation on a Lords-day there would be Bills desiring prayers for about an *hundred sick.* It seized upon all sorts of people that came in the way of it. 'Tis not easy to relate the Trouble and Sorrow that poor *Boston* has felt by this *Epidemical Contagion.* But we hope it will be pretty nigh Extinguished, by that time twelve-month when it first began to Spread. It now unhappily spreads in several other places, among which our Garrisons in the *East* are to be reckoned some of the Sufferers.

These two paragraphs, limited and almost trivial as they are in contrast with contemporary performances, demonstrate some of the concepts of science writing that still exist more than two and three-quarter centuries later. To cite a few:

1. They dealt with public health and medicine. Since people

are always interested in people, especially themselves, science writers long have recognized that reporting about health matters and medicine guarantees a built-in attraction for a majority of mass media readers. Until the space spectaculars of the past decade provided the synthetic thrills of vicarious adventure, readership surveys and content analyses repeatedly showed that medical–health news had the greatest popularity.

2. They stressed the local angle — in this case, smallpox in Boston. Newsmen long have known of this attraction and catered to it. In fact, when the local angle is missing, one well-known science editor refers to it as the "N.I.H. (Not Invented Here) factor."

3. They emphasized "progress," which has almost become a staple in efforts to gain attention for science news. Harris compared favorably (in terms of reduced death tolls) the recent smallpox epidemic with one 12 years earlier — just as today's correspondents at Cape Kennedy or the Manned Spacecraft Center at Houston write about increased payloads, mounting tons of lift-off power, or ever longer "walks in space."

4. They were tied in a vague, tangential way to the military. Science writers often tack some of their reporting to the current interest in Cold War developments and thus hope to benefit from a two-bang firecracker appeal to attract readers for their news items.

Although Harris' two paragraphs were tinted with the editorialization so common among colonial journalists, he made little other effort to interpret the facts; this backgrounding approach is a great distinctive quality of professional reporting of modern science for the mass media audiences. Its absence probably reflected Harris' lack of expertise in both science and journalism as well as his sense of caution engendered by imprisonment earlier in England for offending the governing Establishment. It was the same fear of imprisonment that ended his publication after a single issue: the Royal Governor was offended by some of his non-science articles.

Any survey of science reporting in United States newspapers, no matter how cursory, will illustrate an almost limitless variety in the motivations for such journalistic activities. Since their very name implies a responsibility to inform the public about current developments in various fields, the nation's newspapers have been a channel for reporting those activities in science that are considered interesting, important, or useful, to the general reader. The primary job was to report news. Science, like all the other information sources, has been mined to provide human interest materials, vicarious thrills, and amusing anecdotes. Some news items have been exploited for propaganda or promotion. Still others have contributed to the laymen's education and background needed for informed decisions on public policy.

Basically, a newsman may use the traditional police-reporter approach to science; there are some differences, however. Stories that fall in this category include descriptions and details that a competent, inquisitive correspondent would gather at the scene of a fire, a bank robbery, a missile launching, or a scientific convention.

But to do a truly competent job requires more than a newsman's nose for news. A science reporter has to be alert enough — and this implies an understanding of the basics of "pure" science, technology, and medicine — to ask intelligent questions and to comprehend the answers. He has to understand enough of the jargon of scientists to follow the ideas being discussed. Just as lawyers have their legal terminology, so scientists have a vocabulary of their own — although some editors and publishers still believe this to be a trait more peculiar to the men of science. Every science writer talking to a Nobel Laureate does not have to have an advanced degree in that scientific subject. If that were generally true, there would be few newsstaff members qualified to interview scientists — or, for that matter, graduate students in political science to talk to politicians, or agronomists to speak with farmers and agricultural agents. When newsmen have extensive specialized training, there is some danger that they will assume that all potential readers have their own mental cargoes and so type out stories incomprehensible to all except the specialists. This assumption of reader expertise has happened just often enough to illustrate the point that a journalistic (although not a scientific) catastrophe may possibly occur from a reporter's "knowing too much," as his copy desk might say.

The well-prepared science reporter also must know how to write — in the language of the man in the street. The late Raymond Clapper, Washington bureau chief for United Press, used to insist on political news written for the "milkman in Omaha." His advice for national politics is just as valid in writing a science story for the general reader. Despite the rise in science literacy since World War II, the public's knowledge still falls far behind the specialist's as he discloses his recent experiments on a scientific frontier.

In the early nineteenth century, newspaper readers learned of the Lewis and Clark expedition into the Western territories of the United States just as contemporary readers are informed of the astronauts' space adventures, but the time lag of months contrasts markedly with our contemporary "instant journalism." A century ago, Alexander Graham Bell's demonstration of a telephone call between Boston and Cambridge was reported in the form of a transcript — much along the lines of contemporary press conference quotations of questions and answers from a doctor's discussion of presidential health or from a space specialist's evaluation of the chances for sighting unidentified flying objects from other worlds.

Another approach to science news involves human interest, that attempt to personalize scientists and the work they do. While this type of reporting tends to bother some scientists because they believe it is (and, of course, *it is*) unscientific by their specialized standards, it remains a bread-and-butter item for politicians and others who seek the public's favor and support. And if science really is a study of adaptation to environment, then some of its complaining practitioners might reconsider their own key assumptions and adapt them to their public relations.

Dr. Warren Weaver, that practical philosopher of science,

foresaw possible benefits of human interest features when he wrote that it was necessary to portray scientists as people, as of course they are, and not as "special creatures." Such backgrounding would tend to quiet those persistent superstitions which say that science cannot be understood by ordinary mankind but only by those wonderful, yet strange, "scientific priests." Weaver predicted that with dissemination and general acceptance of such enlightened attitudes, the public — at least many of its members — would cease to view scientists one-third of the time as amusing but beneficial eccentrics, one-third as sorcerers, and the other third as irresponsible rascals.

At least one science popularizer speculated that mass awareness (but obviously not understanding or real appreciation) of the theory of relativity was not unrelated to the portrayal of Albert Einstein in human interest articles as a man who played a fiddle and had gray hair that stood out as if he were charged with electricity. Such portraits of scientists sometimes veer off into making them appear as oddities; this generally does real damage to building an image of the man or woman as he or she is — but it does aid the journalist by giving him a convenient handle. Many scientists, and some reporters, argue that they would rather not have the space and time if it means filtering impressions through this distortion.

Walter Sullivan, science editor of *The New York Times,* once defended the responsible light-touch stories on the grounds of journalistic economics when he explained:

> But most newspapers, even perhaps my own, would soon go out of business, if they only ran stories of world-shaping significance. We must entertain as well as inform, and it's great fun doing so. Of course, the stories that we print should not be frivolous or in bad taste. . . . What a good newspaper should serve is a balanced diet with plenty of meat and a bit of light, skillfully done pastry.

However, even in the 1960's, some scientists failed to accept a need for emphasizing the human interest aspects of their activities. In 1962, the National Association of Science Writers issued *A Handbook for Press Arrangements at Scientific Meetings* which included these paragraphs:

> Many scientists to this day question the "popularization" not of research but of people as well as their research. They cannot accept the notion that the people are often just as important to the science writer and as interesting to the reader as research findings. Scientists, as news-makers, are no different than statesmen, politicians, industrialists and others.
>
> There is nothing undignified or unscientific about a description of a man or woman, his hobbies, family, extra-curricular interests, political, social or ethical viewpoints on human affairs. All this can be, and usually is, written accurately and in good taste. . . .
>
> The scientist can set the tone for an interview of a "personal" rather than "professional" nature by his answers; he need not respond to queries which are in poor taste, stupid or obviously intended to embarrass or harass.

In candor, however, I must admit that some science writers in recent years have wavered between the "For he's a jolly good fellow" approach and the "Dr. Frankenstein" touch. The first applied especially to astronauts, both United States and Soviet. For documentation, simply recall the stories about their wives and children that piled up column inches and filled in otherwise "dead" air time during manned space flights. Transcripts of telephone conversations of wives talking to husbands blossomed in both newspapers and television during the early multi-day flights. The "Dr. Frankenstein" attitude is just the opposite: a commentator plays on stereotypes as he trots out the terror-inspiring machine manufactured by the fictional character who is a scientist of sorts.

The ambivalence about the social impact of most scientific discoveries and applications — whether they are beneficial or destructive — can be well illustrated. To mention just a few examples, consider nuclear energy, which can be used as a tremendous power source for electricity or for hydrogen bombs; the automobile, which is an indispensable American transportation convenience as well as an agency for the annual traffic death toll of approximately 50,000 persons; or the modern jet aircraft, which can carry hundreds of passengers or tons of bombs across oceans in a few hours. The feelings that were aroused by details of bomb destruction at Hiroshima and Nagasaki more than two decades ago prepared the way for some of the emotional reactions that even the beneficial applications of science have not entirely eradicated. So a science writer can point his journalistic compass in either direction and be sure of arousing some interest.

Reporters' efforts to write humorous and "Gee whiz!" stories are viewed by scientists, engineers, technicians, and doctors with even more trepidation than are the attempts to personalize their work.

Although it does not happen often, there are still some crude journalistic attempts to poke fun at some longish term that is part of the stock-in-trade of research. Scientists have built their own terminology, in part, because they require a linguistic accuracy and purity — just as they need a scientific accuracy — that is more demanding than laymen require. No M.D. would think of telling a patient, "Take some of that pinkish medicine in the yellow bottle on the second shelf of the bathroom cabinet." Yet a husband might understand this remark from his wife without any difficulty. Just the fact that a word has many syllables does not provide ground for amusement. On the other hand, as Sullivan pointed out, there are some light, amusing incidents about scientific research, technology, and medicine that reporters should be permitted to tell. There is no reason for scientists to insist that they are all long-faced geniuses who never indulge in horseplay; some of them certainly are not and what they do provides good reading.

Unlike the humor story, the true "Gee whiz!" story is supposed to engender such amazement that readers will exclaim, "Gee whiz! Isn't that something!" Yellow or sensational jour-

nalism's stress on its type of "Gee whiz!" reporting during the
1890's so shocked and intimidated the scientific community of
that day that it was at least a full generation before all but the
bravest scientists would permit themselves to be interviewed
by newsmen. Even in the middle of the twentieth century, the
occasional use of traumatic headlines still had some impact on
young scientists with little personal experiences from which to
draw more realistic judgments. The reactions to yellow journalism were not unjustified. Researchers and physicians had good
cause to fear reporters when their stories might be headlined
as were these three in William Randolph Hearst's *New York
Journal* during 1898:

IS THE EARTH FLATTENING OUT?

A MAN WITH A STOMACH WHO LIVES WITHOUT EATING

A WOMAN WITH NO STOMACH WHO EATS WITHOUT DIGESTING

When such flamboyant reporting appeared in more recent
years, some of the authors were not allowed to go unnoticed.
When a New York City paper prominently displayed a story
with a headline something like: "Science Finds Man Descends
from Horse," rival reporters at an American Association for the
Advancement of Science convention sent this telegram to the
reporter's editor:

Respectfully want to know from which end of horse did
man descend?

In still another presentation of science development, writers
come close to the style of the classroom teacher. This approach,
which one newspaper science editor named "educational fall-
out," has expanded extensively during the 1960's. In 1965, for
instance, an informal survey by the Council for the Advancement of Science Writing, Inc., found at least 14 weekly science

pages in United States newspapers. All except three of these were developed after the start of the decade.

Since 1963, for example, the *Minneapolis Morning Tribune* has carried a Monday morning science page designed for use in secondary schools and written by scientists and newsmen. The *Tribune* staff reports that more than 2,500 classrooms in the Upper Midwestern states use the pages, which appear for 25 or 26 weeks during the school year. An entire page, with illustrations, often printed in several colors, is used. The *Tribune's* brochure for the 1964–65 school year series explained that the articles were "designed to supplement the study of science in Upper Midwest secondary schools by linking class-room studies with today's scientific research and programs." The paper hoped to "provide students with current informa-tion on the theories and discoveries that form the basis of pres-ent-day scientific thought and are fundamental to the general understanding of where we are and where science may lead us in the years ahead." In December, 1966, the *Minneapolis Morn-ing Tribune* received a special George Westinghouse citation from the American Association for the Advancement of Science for the project.

When the *Chicago Daily News* began its science page in 1964, it promised:

> Along with the spectaculars will be stories about those who work at science as an intellectual adventure and create the pool of basic knowledge which nourishes prog-ress.

These science pages come close to being weekly chapters to supplement classroom text and to breaking rather drastically with the convention that newspapers must print what happened in the previous day or the previous week — and little else. It would be naïve to avoid mentioning that these weekly science pages carry considerable promotional value, but there seems to be little evidence that this was their exclusive or even their primary motivation. The emphasis was reversed a century ago when James Gordon Bennett, Jr., sent Henry M. Stanley into what was then "darkest Africa" to find the British missionary David Livingstone. The expedition provided the *New York Herald* with exciting articles that gained world attention and boosted circulation.

In addition to writing for the science pages, some of which are designed specifically for classroom use, many science reporters have been pushed and shoved into using background information that could pass as educating the readers. This may be done through explanatory paragraphs in otherwise straight news items or through entire articles of interpretation and news analysis. Basic research has pushed into areas so totally unfamiliar to the public that the innovating concepts have to be explained in order for the latest development to make much sense. If a science correspondent is going to discuss this work meaningfully, he is forced to educate his audience as he goes along; there is no previous source for such information. So newsmen write about the family relationship of a new "strange particle" in an announcement from Brookhaven National Laboratory just as their colleagues explain the family tree of a new British royal prince or princess.

In discussing this point with physicians some years before these changes had gained wide acceptance, one nationally syndicated science writer put it this way:

> You think science writing should be educational for the public, primarily. But no newspaper editor expects to publish science stories because they are primarily educational. They are published because they are news. If we can educate the public along the way, we're doing well. We can't always do it, and we can't do it primarily to educate. That is the point where we differ. . . .
>
> If I wanted to teach, I would be in a college teaching. I don't want to teach. I want to report news. If I can do that in a way that will be useful, by telling people things they need to know, that's all right. But whether they learn

or whether they don't learn, I am not a teacher. I am not
trying to make them learn a lesson.*

Quite a few other science writers still feel that way, but their
numbers seem to be declining — too slowly, some scientists
might claim. Magazine articles and television documentaries in
recent years have turned to science news, just as some metropolitan dailies have established weekly science pages. When
any one of these three media have time or space, it can probe,
explain, and display, science in depth, and in the process all
three contribute to the education, as well as the informing, of
the general public.

But recent efforts do not satisfy all the people who read and
watch. For example, a Rutgers University professor wrote in
Science (June 8, 1962) after Lieut. Col. John H. Glenn's orbital
flight:

> Television did a remarkably competent job of present-
> ing the actual flight and the festivities that followed its
> successful completion. But during this whole period not
> so much as a single half-hour segment of television time
> on any station or network was devoted to any explanation
> of the scientific background of this exploit. Presentation
> of a few basic principles — such as the concept of an
> orbit, weightlessness, physical conditions in space, and
> the physiology of space flight — could have lent meaning
> and substance to this great technological achievement.

Discussing the Gemini-4 flight of June 3–7, 1965, Edwin Diamond, author of *The Rise and Fall of the Space Age* and a
Newsweek senior editor who formerly was that magazine's
science reporter, said in the Summer, 1965, issue of the *Columbia Journalism Review* that space and television are "ideally

* For a fuller discussion see Hillier Krieghbaum, ed., *When Doctors
Meet Reporters* (New York: New York University Press, 1957), pp. 65–
72.

suited to each other." With a knowledgeable network anchor-man, he said, it is possible to convey a sense of the intricacies of the flight without losing the drama. Then Diamond commented:

> The manned space flight program of the United States provides television with an opportunity to convey an enormous amount of popular science to a mass audience, to inform as well as to divert the viewer. The ability to understand the mathematics of orbital mechanics is only one small part of the space program that requires elaboration for the audience. A mission like McDivitt's and White's encompasses a cram course in science — the medical and physiological effects of space flight on the human body, the physics of the near space environment around earth, the principles of radio communication and of computer technology, the chemistry of gases and liquids. But the television coverage of the Gemini flights, and of the Mercury flights before them, did not so much explain the principles of Newton and Kepler as list them. . . .
>
> Radio-television journalism still seems about a generation behind the newspapers and magazines in science reporting. It still depends too much on cliché and the "oh-it's-all-too-complex-for-you-and-me" style of popular science writing of old (a nucleic acid isn't just called deoxyribonucleic acid; it must have "the jaw-breaking name of . . .").

For later technological spectaculars in space, television and many publications have sought to prepare the general public for possible findings that might result from such activities as a rendezvous in space, an instrument landing on the moon's surface, and a satellite fly-by of and landings on Mars or Venus. *The New York Times*, that showpiece of American journalism, has repeatedly printed glossaries of space terminology and almost textbook elucidations of basic principles involved. Other metropolitan dailies have done likewise, but competition for news space in smaller papers and for time on brief radio and television news summaries frequently has crowded out such backgrounding. Almost without exception, smaller-sized papers and shorter newscasts have all but given up on reporting the stories of advances in such areas as molecular biology, high-energy physics, and oceanography, which certainly have had their surprises but to date have lacked pictorial equivalents of rocket firings.

However, some measure of the increasing sophistication of news coverage in metropolitan dailies is shown by the two glossaries of terms printed in *The New York Times* for the flight of Lieut. Col. John H. Glenn, Jr. (February 21, 1962), and for the walk in space of Major Edward White during the Gemini-4 flight (June 8, 1965). For the earlier flight, the glossary of 38 terms included: Apogee, gantry, go ("Ready to go, fine, in working order, A-Okey."), G-forces, hold, scrub, trajectory, and weightlessness. For the Gemini-4 flight, the 17-word terminology included: Drogue chute, heat shield, microwave, re-entry, and retro rockets. For the Gemini-5 flight, the *Times* printed no glossary of space terms during the entire eight days and chose rather to define words and phrases within news stories.

To illustrate the near-textbook science writing in the mass media, *The Washington Post* of August 22, 1965, included the following explanation when the Gemini-5 fuel cells misbehaved:

> The primary reason that both the hydrogen and the oxygen — necessary chemicals for the fuel cell system — are kept in a liquid state aboard the Gemini is a lack of room to take them aloft in a gaseous state.
>
> Nonetheless, the fuel cell, itself, will operate only on gaseous oxygen and gaseous hydrogen. Hence the need to convert the liquids to gases and to convert enough of both to meet the electrical demands of the Gemini 5 flight.
>
> Though the fuel cell concept is old, the actual com-

mercial use of these chemical batteries is relatively recent. The fuel cell on Gemini 5, for example, has never
before been flown in space.

Essentially, a fuel cell converts a fuel, in this case hydrogen and an oxidant, in this case oxygen, into electricity, heat and water. A fuel cell, unlike a conventional
battery, does not store energy. The fuel cell converts
energy and for this reason it does not require recharging.

This kind of front page journalism comes close to that in a
well written science textbook.*

To see how massive coverage may yield science news with
a variety of oblique approaches, let us look at the coverage of
Lieut. Col. John H. Glenn's flight as reported in *The New
York Times* of February 21, 1962, which printed a staggering
1,374 column inches of news stories, maps, charts, and pictures, related to the space voyage. This meant slightly more
than eight full pages out of an issue of 92 pages. Add the 761
column inches of advertising tied in directly with the colonel's
trip, and the total goes up to 2,135 column inches. This all-
inclusive figure is more than one-eighth of all the news and
advertising space in that day's issue.

News coverage broke down as follows: News text — 962
column inches; pictures, diagrams, maps — 412 column inches.

What was included in this mass of type and cuts?

The main story from Cape Canaveral (now Cape Kennedy)
that dropped off the eight-column front-page banner headlines
ran for 96 column inches on the front and inside pages. It was
a combination of highly competent police-reporter descriptions, human interest quotations ("It was hot in there"), plus
scientific statistics and explanations. Partly science education,
yes, but other things, too!

The Times science news staff did its usual good job. In-

* For fuller discussion see Hillier Krieghbaum, "Two Gemini Space
Flights in Two Metropolitan Dailies," *Journalism Quarterly* (Winter,
1966) 120–21.

cluded were lengthy news items on the global radio and radar tracking network, the missile's electronic "brain," and a 17-inch "Glossary of Space Terms" that should have found its way into hundreds of high school and college science class-rooms.

Approximately a full page of type (175 column inches) re-corded a transcript of most of the messages from Col. Glenn and flight descriptions by Col. John H. Powers, then informa-tion officer with the Mercury Project.

All of these helped to inform the public and, along the way, science education was conveyed to those who read.

But other items had little to do with basic science or space technology. Here are some of those stories:

A short item from London that "Flying John," the hunch favorite, came in second at Wolverhampton Race Track in England.

Seven and a half inches reporting Col. Glenn had earned $245 extra in flight pay for the time he was aloft.

A short on the nomination of the colonel's daughter as presi-dent of her ninth grade class.

The complete text of President John F. Kennedy's con-gratulatory conversation with Col. Glenn.

Twenty-five inches, with illustration, on the Project Mer-cury commemorative postage stamp.

Nine and a half inches on the Wall Street reaction to the flight.

Twenty-four inches of Columnist Jack Gould's evaluations of television and radio performance in regard to the Glenn production.

Some might say that these news items were not science stories and they would be entirely correct. But it was Col. Glenn's flight that triggered copy editors to open their news columns for more truly scientific developments as well as these others.

This is not a plea that these other items should have been spiked by *The Times'* copy desk or that the slugs of type should have been thrown into the "hell box" as useless. I am simply trying to show that the staggering total of column inches on the Glenn flight did not all qualify as "hard news" reporting of science. (The late Dr. Frank Luther Mott, former dean of the Missouri School of Journalism, made the valuable distinction between "hard news" — significantly important news of situations and events which are not exciting for the casual reader — and "soft news" — the obviously interesting news of immediate reward.) Possibly science reporters should be happy that other news writers at last were paying some attention to what impact science and technology were having outside the laboratory and university campus.

1444372

3

THE CLASH OF THE
CULTURES

During the past decade, especially since C. P. Snow's Rede
Lecture in 1959, a great deal has been heard about "the two
cultures" and, as he phrased it, "a gulf of mutual incompre-
hension" between them. While there have been differences
over the estimates of the width of that gap and questions
raised about a third, a fourth, or even a fifth, culture, at least
two varying points of view have clashed in naked conflict
when scientists and physicians have met newsmen. The science
press room has, on occasion, become a cultural battleground,
despite the polite bows before the combatants hurled their
barbs.

Both scientists and reporters have their own professional standards and guidelines for conduct, and each set differs widely from the other. When any group looks at the world through its own colored glasses, troubles and tensions are apt to develop. This has been true at some scientific gatherings covered by newsmen.

Criticisms have been lobbed back and forth with all the vigor of tennis balls at an international competition. Consider the following, which are fairly typical of remarks not infrequently made in conversations and in print.

A well-known physician: "My chief objection to popular scientific articles is that they can't be both popular and scientific. This is understandable when you realize that newspapers and magazines are essentially a part of the entertainment industry. For a science article to be popular it must entertain."

The public relations chief of a national scientific council: "It pains me to say so, but relatively few scientists are yet aware of the enormous importance of reasonable competence in their use of English."

An editorial in an internationally distributed scientific journal: "The material which is printed [in most of the mass media] is usually gee-whiz, Buck Rogers distortions of the facts."

A science reporter with years of experience: "Very often we meet some resistance from the scientists and the doctors, and when we start asking questions, they get on the defensive. . . . They are not making, in short, the same kind of effort to understand our problem as we are trying to do for the American public through newspapers."

Taking a broader perspective, Nathan S. Haseltine, medical writer for *The Washington Post*, winner of the 1953 American Association for the Advancement of Science-George Westinghouse Science Writing Award, and recipient of the 1964 James

T. Grady Award of the American Chemical Society, has described accurately and in detail the difficulties that may arise when newsmen and doctors get together. In the mid-1950's, during a series of uninhibited dinner discussions sponsored by the Josiah Macy, Jr. Foundation,* Haseltine outlined the situation as follows:

> Newspapermen and physicians live in their own worlds. They see the same things, but each views them from his own training.
>
> The newspaperman looks at something new as news, something competitive which he must write about before his fellow newspapermen. If it is a promising medical development which doesn't pan out, that's all right, too. The stories of failures are the pegs on which he hangs later newspaper stories.
>
> When the doctor sees something new — except perhaps when he has developed it himself — he sees something that is yet to be proved. He may see its promises, but until they have been fulfilled he is leery of such claims.
>
> The newspaperman looks at news as something the public has the right to know. And, by God, no doctors are going to smother that right.
>
> The doctor has the duty to protect his patients — and himself.
>
> Codes of ethics for newspapermen are about as binding as each newspaperman's own code of honor. The written codes are younger than most of the newspapermen who are covered by them. The unwritten code of journalism is pliable to every justification made when breaking it.
>
> Physicians are bound by an oath, and a protective code of ethics. The violation of either brings down the wrath, and retaliations, of colleagues. The violating physician's

* For a detailed report of these meetings see Hillier Krieghbaum, ed., *When Doctors Meet Reporters.*

reputation and income may suffer; in fact his very right to practice may be taken away from him.

In this world of freedom of the press, guaranteed by our Constitution, neither newspapermen nor their newspapers are licensed. The newspaperman and his newspaper that break the ethical code are not penalized, other than by loss of circulation when readers turn from the paper in distrust.

All this is neither praise nor condemnation of either profession. It just shows that the concepts and practices of medicine and of newspaper operations are as foreign to each other as the Eskimos and the Hottentots. The doctors and the news reporters go their separate ways, each wondering why the other is so strange.

Most of what Haseltine said about the print media also applies to radio and television, although both operate under federal license "in the public interest, convenience, and necessity." His remarks also apply to Ph.D.'s as well as M.D.'s, despite the fact that scientists — except for a few specialists — are not licensed by the government.

Newsmen are compelled to gather such information and news as they can for publication and, for most of them, their professional integrity demands that they do the best they can under the circumstances. Responsible reporters will try to check their facts to see that they reflect correctly the gist of a scientist's findings, but they will not feel it important to include all possible conflicting or complementary theories or even to credit the contributions of a scientist's predecessors.

If there really is a cultural gap, one place where it becomes most apparent is in the different degrees of reportorial accuracy required for publication in a professional journal and in the mass media. This difference might be called a contest of scientific accuracy versus headline accuracy.

To meet adequately the demands of scientific accuracy requires sufficient details to permit duplication of the reported experimental results by professionals. Enough material is

needed to replicate the completed experiment — and enough
to permit making a scientific judgment on the basis of the
validity of the original results. These procedures have the
added value of putting to a disadvantage the "operators" or
publicity-seekers.

For headline or journalistic accuracy, what is required is
a correct impression or over-all picture of what the scientific
findings mean to the non-scientists. A college English professor does not rush out to a linear accelerator to check some
report on pi mesons. Nor is an accountant going to demand
time off so that he can examine under a microscope the latest
sample of oceanic core showing the composition near the
Mohorovicic discontinuity.

What science news "consumers" want from the mass media
is a bit of the essence of the experiment, not its detailed nuts
and bolts. This relatively simple idea is not understood by
many non-journalists, including some scientists. In the mid-
1950's, for instance, when I asked a random sample of individuals listed in *American Men of Science* some questions
about how to improve popularization of science information
through the mass media, approximately 5 per cent complained
that their chief objection to general news reporting of science
and medicine was that it provided too little information for
them to reproduce the experiments. They had missed the whole
point of the distinction between the technical and scientific
publications and those that served the public audience. Reporting science, technology, and medicine for the popular
media is not to be confused with technical writing or reporting
to the scientist's confreres in his own or a related field.

When Albert Einstein first developed his theory of relativity,
some claimed that it was truly comprehensible to only a few
hundred people in the world. Correspondents tried at the
time to present a generally accurate picture of the implications
of his theory. That they at least partially succeeded is shown
by the incorporation of Einstein's philosophical contributions
into contemporary thinking, although his more abstract details
still elude most people, including some scientists.

The touchy situation between researchers and reporters was well illustrated in the early 1960's when a series of differences developed over newspaper publication of scientific findings before they had been printed in professional journals. Reporting to one's colleagues first has traditionally been considered a part of the conventional procedure. Many scientists view the whole process of pre-publication review by their peers as a mechanism for safeguarding standards of excellence and as a means of preventing "operators" from circumventing these self-policing methods.

This machinery grew up before public funds played much of a role in financing research and development. Science reporters argued that their clients, the general public, had a right to know what was happening to their tax money. For several years, comments were torrid on both sides and the feud constantly lively.

In 1959, *The New York Times* ran a page one report on an experiment being prepared at Harvard which was to use the newly-discovered phenomenon known as the Mossbauer effect to test Einstein's theories. Faculty members had submitted a report on their proposed experiment to *Physical Review Letters,* and this article was awaiting publication when *The Times* story appeared. The journal editor was, as a *Times* science staff member described it later, "understandably upset" since the editor apparently thought the Harvard men had sought out the reporter to gain publicity for their work. *The Times* reporter insisted, however, in recalling the incident, that this was not the case; the newsman had the basic facts when he first walked into the Cambridge laboratory.

Out of this incident and others like it grew an editoral which was printed in the *Physical Review Letters* (January 1, 1960), stating in part:

> Scientific discoveries are not the proper subject for newspaper scoops and all media of mass communication should have equal opportunity for simultaneous access to the information. In the future, we may reject papers

whose main contents have been published previously in the daily press.

A sister physics publication, *Applied Physics Letters,* followed up in its November 1, 1962, issue with an editorial that pushed the embargo somewhat further:

> Moreover, work described elsewhere, for example, in press releases or in the form of abstracts of contributed papers, prior to scheduled publication in APL, will not be considered.

Robert C. Toth, then with *The New York Times,* questioned an attempted embargo on release of news about space research, specifically the Mariner II results, in comments that appeared in the paper's Western edition (December 12, 1962):

> Should the discoveries be given to the press as soon as available? Or should they, like other scientific results, be given first to the scientific community as some scientists demand?
> The public, whose tax funds financed the experiments, have a stake in the matter.

In the January 18, 1963, issue of *Science,* the AAAS weekly, its editor, Dr. Philip H. Abelson, said that he had been asked to adhere to the positions taken by the physics journals and reported that, while he could not completely agree, he was "generally sympathetic to their stand." Then he pointed out:

> I feel that newspapers and scientific journals are not in serious competition with each other. These media are worlds apart in audience, coverage, and provision of technical detail. . . .
> The alert scientist gives only limited credence to newspaper stories. He finds them valuable as indicators

of important events. To obtain full details and sufficient information to judge the validity of a claim, he knows he must consult the scientific literature.

At a University of Pennsylvania symposium* in October, 1963, Abelson discussed his fear that it was becoming easier for scientific "operators" to get around the conventional self-policing methods of pre-publication review and scrutiny by readers, some of whom are likely to check up on the reported findings. Then he continued:

The publicity-seeker can get his story direct to reporters. Newspapers do not print sufficient technical detail to permit other scientists to confirm the facts behind the announcement. Even if they did and the story were proven false, the "operator" could shrug his shoulders and say he was misquoted. . . .

Using shrewd tactics an "operator" can establish himself as a newsworthy person. This can open the road to research grants and even academic advancement. Only a few scientific specialists will suspect that he is a phony, and they will have no practical mechanism for penalizing him.

Thus a situation has arisen in which intellectual dishonesty can be rewarded ahead of integrity and "operators" need not fear penalties arising from exposure. If this trend continues, a competition in dishonesty may occur, and science news reports will become meaningless. The menace would disappear if journalists would confine their stories to reports based on manuscripts accepted for publication.

At the same Philadelphia meeting, Walter Sullivan of *The New York Times* answered Abelson as follows:

* See *Communications and Medical Research* (Philadelphia: University of Pennsylvania, 1964), pp. 17–18 and 55–56.

We know some of these operators pretty well. We like to think they can't flim-flam us, though they probably do once in a while. There are a few commercial entrepreneurs who flim-flam us once in a while, too, but we do our best to avoid it.

But let us not confuse science operators with those both willing and able to speak for science. Some of them have a very special gift. They can describe in layman's terms, with eloquence, style and sometimes drama, the battles being fought on the frontiers of science. In some cases they are themselves participating in these battles, important and effective participants. . . . Thank goodness we have them, for otherwise the public understanding of science would be far poorer.

Toth and some of the other Washington science newsmen queried the National Aeronautics and Space Administration about how it proposed to distribute newsworthy information on future satellite activities, suggesting that such reports be given to the mass media when articles were submitted (not published) in professional journals. NASA came back with a compromise: Most of the scientific reports to be printed in *Science*, which has a Tuesday deadline for a Friday publication date, would be released to reporters when accepted for publication by *Science* editors. NASA officials also agreed not to hold up news on technological accomplishments, such as a picture of the backside of the moon, for prior or simultaneous publication in scientific journals, and said further that discoveries of "transcendental importance," such as confirmation of life on Mars or Venus, would be given to everybody at the same time.

Toth retorted with caution, "I guess we've gotten something of a victory, if it holds up." And science correspondents sought to extend the announced NASA policy to other governmental agencies and institutions.

Suppose that the scientists, rather than the professional

science reporters, wrote up the day-to-day events and occurrences in the world of science for the general public. How
would that be? Both scientists and newsmen agree that in
most cases it would be an impossible mess.

Pierre C. Fraley, former Philadelphia *Evening Bulletin*
science writer and for five years executive secretary of the
Council for the Advancement of Science Writing, Inc., reflected the viewpoints of many reporters when he expressed
his own opinions in the *NASW Newsletter* (March, 1963):

> It is my feeling that very few scientists are either
> equipped to do this or have the inclination or the time to
> do it conscientiously on a continuing basis. Also, there is
> a fallacy that having detailed, technical knowledge about
> a subject automatically confers an ability to communicate
> and be articulate about it.
>
> Should the scientist-science writer be a chemist, a
> physicist, a biologist, a cultural anthropologist? Does
> the fact that a man is a darn good low-temperature
> physicist make him competent to report on the latest de-
> velopment in sociology, or zoology or astronomy? All
> these fields and many more the science writer has to
> cover. . . .
>
> The newspaper science writer has a great advantage
> over the scientist because writing for the public is what
> he does every day and if he is a good science writer he
> knows what the public wants, or needs, to know about
> science and how to present it concisely, clearly and under-
> standably. He is aware of the pressure of time as well as
> limitations of space. He is basically on the side of the
> reader.

Fraley admitted that a core of scientists qualified as "excellent science writers" have contributed enormously to the public understanding of science. But he pointed out that most of
them were known for their books and occasional magazine
articles or for their radio and television guest appearances.

Not many, he explained, had written directly for newspapers, covering spot science news on a day-by-day basis.

Then he summed up his attitudes:

> The professional science writer knows that the job of interpreting and reporting science to the public is as overwhelming as the explosion of scientific information and that as many articulate people as possible — scientists and humanists — are needed to keep us abreast of developments.

M. W. Thistle, a scientist who became chief of the Public Relations Office of the National Research Council of Canada, was even more blunt in an article printed in *Science* (April 25, 1958) when he claimed that scientists produced "some of the worst-written documents in the world." He elaborated:

> I know a lot of scientists whom I love, but whose operations in the English tongue remind me of an elephant on stilts — ponderously inelegant. . . .
>
> Aside from their need for reasonably adequate English as an essential part of their written and spoken reports to one another, they will need English for communicating with laymen at various levels — for example, in the trade journal article, the "popular" speech, and the interview with the press. Some of them do these things very well indeed; but far too many of them do these things rather badly.
>
> The ones who do it badly err on two counts: a bumbling, fumbling use of the language itself and a thoroughly mistaken idea of how much detail is required. There is no substitute for adequate training in writing and speaking. . . .

Dr. Frank Fremont-Smith has spent much of his adult life with the problems of intercommunications across disciplinary boundaries, first with the Josiah Macy, Jr. Foundation and later with the American Institute of Biological Sciences. He made this observation:

> On the medical side, I would like to say that we need desperately, very desperately indeed, the good will and the good work of the science writers. It seems to me that the medical profession, the universities, and hospitals have ignored too long the fact that they can be successful only with genuine public support and they are going to get genuine public support only if their story, their very dramatic and thrilling story, is appropriately told to the public. There is no better group to tell this to the public in terms that the public can understand — because, God knows, we cannot make ourselves understood to the public — than the intelligent, thoughtful science writers.

It is possibly that there always will be some unhappiness between those who do scientific research and those who report their activities. However, both groups may find a bit of comfort in a remark by Dr. Paul B. Sears, whose scientific credentials are documented by his membership on the National Science Board and his presidency of the American Association for the Advancement of Science, the nation's largest group of scientists, and whose writing activities have provided several best sellers that have helped to initiate public policies for conservation and ecology. Sears said:

> For the vital task of keeping the public informed of the nature and progress of science we must depend upon a group of hardworking craftsmen, the professional science writers.
>
> These individuals have been schooled in the hard discipline of journalism, where every word must count. They do not have the advantage of the ordinary reporter whose knowledge comes to him in the vernacular and by simple observation. They must translate from the most artificial and involved language man has devised, know

what it means and respect that meaning in their translation.

I am certain that where science suffers in the current press, the fault does not always lie with those who report it. I would even extend this encomium to the majority of leg-men of the press whom I have met.

Too few scientists try to exercise the same degree of skill in reporting results that they do in getting them. If each would try, after writing a technical paper, to translate it into simple discourse, he would benefit the public which supports him, the writer who must interpret him, but most of all himself. For clear writing is a prime test of clear thinking.

Shortly before he became director of the National Science Foundation, Dr. Leland J. Haworth, in a speech to the American Physical Society, said that scientists had been remiss in telling the layman about their work, thus preventing him from enjoying "the thrills of better understanding of the fundamental principles of nature and of the impact that science has on his daily life and on world affairs." Then he touched on a typical reaction:

> Often we simply have satisfied ourselves that we have told the citizen of our activities by repeating our own shoptalk and catch phrases in our public appearances and press releases. When asked for further details, we have gratuitously provided copies of our highly condensed and sophisticated technical papers and let the matter drop.

Some scientists, even when they try, find it difficult to understand this "strange breed" of newsmen. It may help them to think of news as a continuing series of reports about what is new — at least new to the reporters and to the general public — and that the audience may have an attention span that can be easily diverted to another new (or news) topic.

Newsmen may find scientists just as difficult to comprehend and they too should exert the effort to understand the philosophy of science, both basic and applied.

Writing in the science writers' own publication, the *NASW Newsletter* (December, 1962), Alton Blakeslee, Associated Press science correspondent and repeatedly an award winner during his several decades with that press association, warned his colleagues that "we must not become the captives of the scientists, or be overinfluenced by scientists' suggestions as to how their work should be presented to the public." Then he continued:

> Some such suggestions are sound, but others are pretentious and ludicrous in the light of the practicalities with which we live.
>
> Our responsibility is to tell other people the fascinating and developing story of science and medicine. We must think of ourselves always as communicators, describing *one* segment of the world of ideas and action. We mustn't allow scientists to convince us we carry some special torch, and deserve special consideration above all other writers and editors.
>
> Secondly, we must avoid becoming regarded as simply another pressure group, with the usual fate of such breast-beaters. It is easy to proclaim that science, and therefore science writers, are mighty important. The prestige we win by what we do in the public interest, individually and as an organization, will be far more forceful. . . .
>
> If we should approach editors and publishers with the attitude they need education and should listen to teacher, the sound of the slamming door should echo through the land.

Obviously, reason and right do not reside solely on either side. Neither has a monopoly on guidance to how the world should be run. Yet, if both scientists and journalists will seek it, there should be much common ground. Mutual back-biting will accomplish little, if anything, of value. There is little justification for criticizing a doctor in Monday-morning-quarterback fashion because he did not anticipate an unexpected drug reaction during a difficult diagnosis, nor is it reasonable to blame the researcher because his basic findings on a chemical were diverted to a pesticide that eventually endangered the public when farmers improperly applied it to food crops. Nor is there valid reason to complain if a newsman's well-researched story proved ineffective when the time came for a sponsor to renew a supporting grant.

Communications falter, Blakeslee pointed out another time, when either writer or scientist assumes that the other should know all those things that it has taken him part of a lifetime and long training to learn. Yet numerous problems can be solved when various groups meet on some solid, common ground. Then Blakeslee added:

Compromises must be reached. They should be reached on the basis of what is best for human beings. If either profession is being selfish, authoritarian, myopic or ignorant in any of its news or demands, it should yield to the other, if the other has more solid grounds.

The issue is human beings. We've made a start toward serving them better. We honor and serve our professions the more by doing exactly that.

4

THE DIFFUSION
OF SCIENCE NEWS

When a caveman first discovered the uses of fire, he undoubtedly told his tribesmen about it in a series of grunts that conveyed a strictly limited meaning. Sir Isaac Newton wrote some reports on his experiments in Latin and delayed publication of others for years, even refusing the perusal of his colleagues in the Royal Society. Today's scientists and engineers, however, inform their fellow workers of their results promptly; most recognize this as an obligation to their sponsors and, hopefully, to the general public. Much of this modern dissemination involves the mass media.

Behind this multiplication of potential audiences remains a

key question: Just how much does the non-scientist public need to know? As pointed out in Chapter 1, one tentative reply would be: enough to keep informed on a vital segment of contemporary culture and thus participate meaningfully in public decisions that are a part of the democratic process.

Agreement on how much science is required for such participation is hard to reach; not even scientists have arrived at a common answer. For instance, C. P. Snow in *The Two Cultures and the Scientific Revolution* argued that an inability to describe the second law of thermodynamics was the scientific equivalent of the confession that one had not read a work by William Shakespeare. Dr. Pendleton Herring, president of the Social Science Research Council, was far less demanding in a 1960 talk to the American Association for the Advancement of Science. And many of the nation's educators working on new science and mathematics curricula for high school and college students would hope their texts encompass a lot more information.

Journalists talk about the public's right and need to know what happens, and almost inevitably their emphasis is on the words "right" and "need." This book will go into the questions of secrecy and censorship in regard to science later, but here I want to focus on the word "know." Like some other popular words and phrases, its meaning varies for different individuals and groups.

Just what does a reporter mean when he says he is writing his story or script so that the public will know about a topic, an idea, or an event? Is it enough for the reader or viewer to reply, when asked, "Oh, yes, I saw something about that but I don't remember what it was"? Is it satisfactory if a reader or viewer retains incorrect ideas and wrong interpretations of what the writer was trying to say? Certainly these are not problems confined to science reporting. There is even some evidence that these problems may be less bothersome with this kind of news than with the many others where blinders of emotion and bias are far more prevalent — despite the fact that the science journalist has the constant problem of translating technical and complicated material for laymen's comprehension.

Degrees of transfer of information fall into at least five categories. It is entirely possible to split several of these, but the basic groupings include:

1. *"Pure" ignorance.* Here, nothing gets through. The reader or listener may not have seen the article or heard the program. Even an ardent reader skips more than he reads in the larger daily papers, the fat magazines, and the deluge of current books; likewise, the avid listener misses many programs that are broadcast — even if his radio or television set is turned on all day. Most individuals tend to forget what does not interest them, even if they read or heard it in the first place. So, when members of this group are queried, they answer that they never saw or heard it.

2. *Awareness.* This involves marginal recall. For instance, a person in this category might reply, "Oh, yes, I read or heard something about that but, for the life of me, I can't tell you any details."

3. *Misinformation.* Here a person distorts what was available or garbles several topics with a mass of inaccuracies. Most of this is unintentional, but not all of it. One illustration of this garbling is the man who was asked several years ago to tell what radioactive fall-out meant to him and who replied, "It's any program that's on the air." When the poll interviewer explained that he meant fall-out from atomic bombs, the respondent insisted, "Radioactivity to me is just what I told you."

4. *Comprehension,* either vague or in-depth. In this category fall both those people who have a casual, but fairly accurate, understanding and those who can recite a full-fledged explanation. The differences are in precision and details.

5. *Action.* Even the well-informed sometimes fail to take the steps that will translate what they know into things that, conceivably, may save their lives. The difference here is in motivation, or, put another way, in the accumulation of knowledge

for its own sake versus the gathering of information for the sake of doing something about it.

A well-constructed public opinion poll usually gives information about these degrees of information transfer. At the time of the first Sputnik's launching, I was chairman of the National Association of Science Writers' Surveys Committee which was engaged in ascertaining attitudes toward science, scientists, and science writing under a grant from the Rockefeller Foundation to New York University.* Part of the questionnaires, prepared by the University of Michigan's Survey Research Center, which did the field work, included this question:

> Have you heard anything about launching a space satellite, sometimes called a man-made moon? From what you've heard, what is the purpose of launching these space satellites?

An initial survey was taken in the spring of 1957, six months before Sputnik's launching; a second was taken a year later or six months after intensive media coverage of both Soviet and United States satellite launchings. Thus these surveys provided a convenient before-and-after basis for measuring media performance — one of the best, if not the best, currently available on science news diffusion. The two surveys showed a massive movement of 45 per cent of the general public from the "Heard nothing" group in 1957 to the "Heard of satellites" category in 1958. This was indeed impressive, but the whole picture showed that even such intensive press, magazine, radio, and television coverage failed to establish a solid foundation of scientific information in three persons out of four (73 per cent) in the entire nationwide sample. Those who "Heard nothing" fell from 54 per cent before Sputnik to 8 per cent

---

* The full statistical results were published in *The Public Impact of Science in the Mass Media* (Ann Arbor: Survey Research Center, University of Michigan, 1958) and in *Satellites, Science and the Public* (Ann Arbor: Survey Research Center, University of Michigan, 1959).

after. Those who were vaguely aware of satellites and recalled hearing "something," but no details, increased from 14 per cent before to 23 per cent after. Two vastly expanded categories after Sputnik were those who thought of satellites in terms of "Competition with Russians" (20 per cent) and "Future possibilities" of space travel (17 per cent).

Those who replied with scientific information were an elite group. Some of the comments were as general as "To collect scientific data" or "To find out more about outer space"; others were as detailed as the comments of a young Florida civil engineer who answered, "To study atmospheric conditions, temperature of outer space which would give some indication of the friction to be expected, and the intensity of cosmic rays. Also to study weather and possibilities of TV stations for military use to work in finding missiles." This literate group rose from 20 per cent before Sputnik to 27 per cent after. A rather modest change compared to a 45 per cent decline in the "Heard nothing" category.

Over longer time periods, however, there may be greater information transfer. At least one pair of public opinion polls which asked, "Would you tell me what is meant by the 'fall-out' of an H-bomb?" seemed to indicate that in over more than half a decade the general public was making more correct approximations. The American Institute of Public Opinion (Gallup poll) asked that question in early 1955 and again in late 1961; this period bracketed considerable public discussion and media reporting of this topic, which had also been a campaign issue in two intervening presidential elections. The figures were:

	REASONABLY CORRECT INFORMATION	INCORRECT INFORMATION OR DID NOT KNOW
APRIL, 1955	17%	83%
DECEMBER, 1961	57	43

Early in 1966, the Columbia Broadcasting System scheduled two national health tests and found that the American public had gapping holes in its knowledge of this topic. But it also discovered concentrations of at least basic information. Compared with scores on a citizenship test, the scores on the CBS health tests showed higher percentages in both "poor" and "excellent" categories. For a comparable science test in April, 1967, results followed the two health broadcasts generally—except for a lower scoring under "excellent." Statistics were:

	COMBINED NATIONAL HEALTH TESTS	SCIENCE TEST	CITIZENSHIP TEST
POOR	48%	47%	34%
FAIR	25	26	47
GOOD	14	22	14
EXCELLENT	13	5	5
	100%	100%	100%

Conclusion: The mass media can arouse public awareness in a relatively short period of time; in the cases of the Sputnik surveys, it was a single year. But a corollary indicates that it is difficult, if not impossible, to develop deep, abstract concepts within such a limited time through the use of the conventional spot news coverage and features; long-range diffusion may be something else again, as witness the fall-out answers.

Perhaps those who want the mass media to take over the educators' role should restudy their assumptions. Teachers are here to stay, certainly until weekly science pages and television documentaries become far more widely used as educational tools than at present. The mass media, however, can create awareness and stimulate those who wish to learn more. Better backgrounding and interpretations of the spot news can supplement the bare bones of today's events and thus contribute to educating while ostensibly informing the public.

Much of this has been confirmed by Dr. Wilbur Schramm of Stanford University who looked into a variety of research studies about the public's knowledge of science and summarized his finding in a 1962 memorandum to E. G. Sherburne, Jr., then Director of Studies of the Public Understanding of Science for the American Association for the Advancement of Science:

> Almost everyone today knows something about science, and recognizes it as an important force in modern life. In general, most non-scientists recognize science without understanding it. To them it is the thing that makes possible such esoteric behavior as space travel, is responsible for the wonderful cures physicians can perform, and is behind practical everyday wonders like automobiles and television. How it does these things is very little understood. Therefore, a certain amount of magic and myth comes to be associated with the research laboratory and the white coat. Furthermore, because a certain amount of suspicion tends to develop toward activities people do not understand, there develops a dangerous mutual suspicion and mutual ignorance between intellectuals which C. P. Snow described in *The Two Cultures*. That is, intellectual life tends to bipolarize between those who understand science and those who understand the traditional culture, and between them the majority of the public who really understand neither.
>
> Unquestionably, the level of scientific knowledge is rising (as may be inferred either from a time series of studies of the lay public, or from studies of scientific knowledge in different age groups). But it is doubtful that this rise in public knowledge is even keeping up with the advance in scientific knowledge among scientists.

Too little is known about how to arouse motivational drives — what Dr. Leona Baumgartner, former New York City Commissioner of Health, called "education-leading-to-action." Possibly better public schooling, reinforced by improved media efforts, is part of the solution; but this still leaves unresolved the question of how to motivate those adults who finished their formal education years ago and who now seem unmoved by news articles and programs. While this question intrigues public health officials and social science researchers, its ramifications are also appealing to science popularizers, including those working through the mass media.

Many people who could have afforded polio shots or could have gotten them free just did not do so. Why? As Dr. Baumgartner told the 1961 National Health Forum:

> We have put much time and work on this, and, goodness knows, there have been bushels of public service announcements in all media, even when they were not hard news. But there are large numbers of people we simply haven't convinced. We ought to admit we just don't know how. Our techniques for breaking through in a situation like this are far less developed than, for example, our techniques for the study of virology. Yet the laboratory product is bound to fall short of full usefulness if we don't learn how to make sure people take advantage of it. We need research in the nature of communication — solid, scientific research of the quality that led to the laboratory discovery.

In November, 1954, a team of University of Michigan researchers sought to determine the effect on people's attitudes and reactions when the mass media covered one of the early studies on a possible relationship between smoking and cancer. The 228 individuals surveyed were chosen by an objective selection technique and were residents of Ann Arbor, Michigan, where students or university employees comprise approximately one-third of the population.

The poll showed that non-smokers, much more frequently

than smokers, reported that they almost always read articles about health and science. When asked whether they believed smoking tended to cause cancer (which was the prevailing but not exclusive viewpoint in the numerous news items), slightly more than a quarter of the smokers (28 per cent) accepted the relationship, whereas more than half (54 per cent) of the non-smokers felt that the relationship had been established. Remember that both groups were reading the same newspapers, listening to the same radio and television programs, and, generally, getting the same magazines. Looking at the dual role of education and smoking versus non-smoking, Dr. Charles F. Cannell and Prof. James C. MacDonald wrote in *Journalism Quarterly* (Summer, 1956):

> For non-smokers, the higher the education the *more* likely the person is to accept as proved this relationship between smoking and cancer. Among the smokers, however, the higher the education, the *less* likely they are to believe the relationship.

Much the same findings were repeated after the U.S. Public Health Service's 1964 report on *Smoking and Health* and other government warnings about the hazards of smoking cigarettes. After a temporary decline, Americans by 1966 were smoking more than before. According to the U.S. Department of Agriculture, total cigarette sales for 1966 were an estimated 542,-000,000,000 or about 2.5 per cent higher than the previous year and more than one-third higher than ten years earlier.

After the release of *Smoking and Health,* Dr. James W. Swinehart, a research associate at the University of Michigan School of Public Health, studied the reactions of a panel of 128 university students. He wanted to ascertain if their views changed as the report lost its novelty and immediacy. Swinehart had the students fill out questionnaires on their beliefs and attitudes two weeks, one month and three months after the Surgeon General had released the report. On all three series of questionnaires, acceptance of the report's conclusions was significantly greater among non-smokers than among smokers but there was a gradual decline in acceptance by both groups as time passed. The total percentage of those who felt "certain" the report's conclusions were "accurate and reasonable" dropped from 46 per cent to 38 to 30. Yet Swinehart found 17 per cent of the non-smokers or smokers who had quit began or resumed the habit while only 15 per cent of smokers who tried to quit or to reduce smoking said they were successful. He reported in the *American Journal of Public Health* (December, 1966):

> To summarize briefly: during the period studied, there was an increase in smoking and a decrease in acceptance of the report, recall of its findings and endorsement of possible government actions with regard to smoking and health.

Conclusion: Individuals wear the colored glasses of their own beliefs and emotions when they pick out news items and they interpret what they have read or heard from the same basis; then they use all their education to reinforce and justify their reactions.

For those who believe that wisdom and knowledge could take over the world if only everyone were educated, there may be a rude awakening ahead in an age of mass higher education. As Dr. Baumgartner accurately commented, techniques of researching communication certainly need more study and improvement.

Another window for viewing what motivates individuals to read science news was opened by the 1957 NASW-NYU survey when respondents were asked to pick from a list of five the reason that came closest to explaining why they read science news. The list and the percentages for first and second choices were:

	FIRST CHOICE	SECOND CHOICE
1. "I like to keep up with things that are going on."	34%	18%
2. "Science may determine whether my family and I, and the world itself, will survive."	18	21
3. "Science is helpful to me in everyday life."	13	15
4. "Science is interesting."	12	18
5. "Science is exciting."	2	3
Claimed to have read no science news	17	17
Not answered, didn't know, no second choice	4	8

In attempting to generalize from these findings, researchers at the Survey Research Center who did the field work for NASW and NYU commented:

First, science is very much a part of the total range of things it is important to know about. Furthermore, science is tied in to concern with the problem of survival of our culture. Finally, interest in science, as such, whether from intellectual or practical reasons, is only part of the motivational picture. The latter point reminds us that interest in science is not based entirely on a narrow pragmatism or on specialized intellectual tastes. True, the more informed and more educated consume more science, and true, the content of recalled science tends to be applied science. But it is also important to keep in mind that broad concerns and interests appear to lie behind this picture. In other words, it would be oversimplification to assume that only very practical science or very esoteric science will serve the "average" and the "elite" science consumer, respectively. It is the way science ties in to a total concern about the environment that seems to dominate the reasons for attention to science.

It would seem to follow that isolated bits and chunks of science news might satisfy some in the science audience, but that the bulk of the group would prefer science-in-context — science news that has meaning because it helps make sense of the world, whether that world is seen as benign or fearful.

Although this report was first distributed in 1958, anything even approaching extensive, "science-in-context" coverage by much of the mass media had to wait until well into the 1960's. True, some magazines have been doing such reporting for a long time, but most of these have been specialized rather than mass circulation publications. Recently, more dailies and some television documentaries have swung over too. Obviously, there remains room for considerable expansion of "science-in-context" reporting.

Various studies of news diffusion show that information about some events seeps out like the traditional molasses in January, while other happenings shoot through the "gates" of the communication process like a missile lifting off for a space probe. The study of most interest to science journalists is one by Drs. Paul J. Deutschmann and Wayne A. Danielson. Published in *Journalism Quarterly* (Summer, 1960), it described how three different college communities — Lansing, Michigan; Madison, Wisconsin; and Palo Alto, California — learned about the launching of Explorer-I in January, 1958.

These three communities represent something close to a communicator's multi-media paradise. Each had television service, local and metropolitan daily newspapers, and a full complement of radio stations. Virtually all households were in the potential media audience: the number of those having radios varied from 90 to 98 per cent; television, from 90 to 92 per cent; and the number of those receiving one or more daily newspapers ranged from 93 to 95 per cent.

The sample in each community was drawn on an "every nth" basis from local telephone directories, and those interviewed were asked by telephone when they had first heard the news and through which medium. The telephone calls were made toward the end of the first 24 hours while recall could be accurate. Thus the study gave a portrait of news diffusion by both time and media.

For the combined statistics, well over a third of the respondents reported they knew about the Explorer launching within the first half hour, and by bedtime of the first evening after the launching (after about three to four hours of diffusion time), approximately half of those questioned in Lansing, Madison, and Palo Alto said they had heard of the event. By late afternoon of the second day, the "knowers" had gone beyond the 90 per cent level; by the time the pollsters stopped late the second night, all of those who answered in Palo Alto had said they were aware of the Explorer flight, and in Lansing and Madison 93 per cent gave comparable responses.

How individuals first learned of the news fluctuated, although television ranked highest in all three communities. The variations existed despite relatively comparable media facilities. The following table shows percentages for initial use of the four media:

	TELEVISION	RADIO	NEWSPAPER	FACE-TO-FACE OR PERSONAL CONVERSATIONS	SAMPLE SIZE
LANSING	40	20	17	23	167
MADISON	36	29	22	13	125
PALO ALTO	61	18	10	10	38

A comparable study of 212 Dallas residents made after the assassination of President John F. Kennedy showed that 83.7 per cent knew of the shooting within the first 30 minutes, 92.3 per cent after an hour, and 94.7 per cent after 90 minutes. Well over half of the Dallas sample (57.1 per cent) first heard of the assassination from another individual, such as a relative or neighbor, a quarter (25.9 per cent) from television, and 17.0 per cent from radio.

When it came to using a medium to reinforce what was

learned from the initial "bulletins" or "news breaks" about the Explorer launching, newspapers were utilized far more than other media. The statistics were: in Palo Alto, 71 per cent used newspapers; in Lansing, 50 per cent; and in Madison, 41 per cent.

Deutschmann and Danielson pointed out:

> We certainly have evidence which suggests that TV — even though primarily an entertainment medium — plays a major role in delivering important news. And we also have evidence which suggests that radio is still doing a big news job. Newspapers tended primarily to supplement the broadcast reports.

Such is the pattern for major news events in science, technology, and medicine that result in the interruptions of broadcasting and the replating of front pages. But these comprise a very small share of what is going on. What about the dissemination of scientific news of lesser value — from the journalist's viewpoint?

To cite the findings of the NASW-NYU survey again: The newspapers remain the workhorse for conveying most science information to most people. This undoubtedly is still true, despite television's advantage during space spectaculars.

In this 1957 public opinion poll, a national sample of 1,919 adults was asked to recall actual science news items from any medium. The Survey Research Center staff pointed out that this approach might result in underestimating the proportion of people who actually read or heard something about science, but the researchers felt that for an individual to be considered truly a part of a science audience he should be able to remember, without prompting, some recent news items. Three out of four persons (75.6 per cent) recalled at least one science story or broadcast from at least one news medium. One-tenth of 1 per cent recalled something about both science and medicine from all four media and, at the other end of the recall spectrum, 24.4 per cent remembered no recent news on either science or medicine.

A breakdown of the figures showed the following:

MEDIUM	RECALLED SCIENCE	RECALLED MEDICINE
NEWSPAPERS	36.6%	60.2%
MAGAZINES	20.5	19.2
TELEVISION	21.5	25.0
RADIO	8.3	7.4
NO RECALL	48.2	31.4

Part of the heavy reliance on daily newspapers may be due to their ability to present a smorgasbord of news — in contrast to broadcasting, which must concentrate only on those items that will appeal to a massive audience. A space flight may be exceptionally well covered by television — in a realistic sense, it takes the viewer to the site of the launching — but, except for these spectaculars and some documentaries, television and, even more, radio tend to by-pass reporting-in-depth on science, technology, and medicine. Since this holds true for many non-dramatic segments of the total news picture, no special discrimination is involved here.

5

WHAT GETS INTO PRINT
AND ON THE AIR

"The most striking new element was science."

This sentence in the *Columbia Journalism Review* (Summer, 1962) summarized a comparison of a selective sampling of United States newspapers dated January 10–11, 1947 and January 18–19, 1962; the dates were those of the President's annual budget message to Congress. As the *Review* added, "There was thus assured one comparable national story, but not one that would occupy the entire front page." As it turned out, both days also had "strong science-related" news: isolation of the polio virus in 1947 and preparations for a moon shot in 1962.

Of the 1947 pages reproduced in the *Review*, three of the ten gave front page attention to the polio virus, while 15 years later five displayed the space flight preparations; two other papers which had passed up a front page display of the moon trip had other science stories there.

Only one newspaper, the *Minneapolis Morning Tribune*, played up both the polio virus story and space story on the front page. Victor Cohn, one of the country's outstanding science reporters, signed both *Tribune* stories. Both *The New York Times*, which had a single column item on its 1947 front page, and the *San Francisco Chronicle*, which had a six column, front page feature on the Stanford University polio discovery, passed up Cape Canaveral news on their front pages 15 years later. The *Chronicle* was one of the papers featuring another science item, but *The Times*, with the largest battery of science writers among United States dailies, by-passed front page science news on that date.

The *Review's* findings illustrate, in miniature, how little we do know about science coverage during recent decades. Students and commentators on journalistic affairs have only infrequently put inquisitive fingers into the news stream to test its contents and so there is no continuing, consistent measurement of what science developments have gotten into print or on the air — and what has been omitted. The over-all trend has been toward increasing display but it has been an irregular performance.

However, some broad, sweeping shifts are clearly obvious. Three times during the twentieth century, events have triggered spectacular jumps in science coverage by the print media, with the last one also causing quite a stir in the younger medium of broadcasting. The first big jump came during and just after World War I when chemists were able to synthesize products to replace many of those cut off by the German U-boat blockade and to battle "gas" warfare. The second came during World War II when physicists, engineers, and technicians helped to develop radar and long-range aircraft and particularly when they harnessed nuclear energy for an atomic bomb. The third followed the launching of the first Soviet Sputnik into orbit; with their overtones of cold war competition, these stories have remained front page material during the early years of the Space Age.

The most extensive mass of repeated surveys on science news reading is offered in "The Continuing Study of Newspaper Reading," a series of readership surveys of 130 United States dailies from 1939 to 1950 conducted by the Advertising Research Foundation, Inc.; the Foundation is associated with the American Newspaper Publishers Association, a professional organization which includes representatives from most of the country's daily papers. Data were based on personal interviews with approximately 50,000 individuals. Sample newspapers varied from circulations of 8,570 to 635,346 and were from every geographic region of the nation. Circulation of the 130 newspapers totaled 11,107,379 as of the dates when the surveys were conducted. Dr. Charles E. Swanson, then research professor at the University of Illinois, studied the 40,158 news-editorial-feature content — "whatever content did not seem to be advertising," as he said — and reported his findings in the *Journalism Quarterly* (Fall, 1955). Among his 40 categories of story types were "Science-invention" and "Health-safety."

Swanson found the following facts about science and science-oriented news items:

1. Among the 40 categories, the two related to science were, even when combined, of only minor importance among the 40,158 items analyzed. "Health-safety" made up 1.1 per cent of the total with not quite 450 stories, while "Science-invention" made up 0.6 per cent with approximately 250 stories. This means that the two topics together accounted for less than one out of every 50 items printed — and there were not enough stories in the "Science-invention" group to average even two items in each of the 130 papers studied.

In contrast, sports accounted for 11.6 per cent of the total items, while a grouping called "Social relations," which included the traditional social notices, ran slightly behind with 11.4 per cent. Comics, war news, human interest stories, business and industrial news — each totaled more than *double* the number of items given to science and health *combined*.

An obvious conclusion to be made from this continuing study of 130 dailies taken before, during, and after World War II is that "Science-health" news got far below the smash play found in more recent daily papers after a missile launching or an evaluation of a polio vaccine.

2. Despite the strictly limited number of science-oriented items published, those that got into print attracted considerable attention. Both these categories ranked somewhat above the average readership score for all items in the 40 groups. Swanson found that the average readership for all news items was 20.2 per cent, about one reader out of every five; for "Science-invention" the figure was 21.8 per cent, and for "Health-safety," 20.7 per cent.

While percentages give some raw clues, possibly more illuminating for us are the topics that ranked ahead of and behind these related to science. Comics were the far-away favorite, being the only category read by a majority of all those questioned in the surveys. Among the others that led the science groups were stories about war, defense, fire and disaster, human interest, weather, major crimes, and those that had some social significance. Newsmen have long known that people — almost without exception — are interested in their fellows and in catastrophes. In fact, Dr. H. A. Overstreet once criticized the mass media for what he called their "vested interest in catastrophe." But topics that ranked lower than these two science-oriented groups are revealing too. Among these were items dealing with governmental happenings (local, national, and international), labor, politics, education, business-industry, society, sports, and taxes. Some of these (such as business-industry, education, politics, society, and sports) got more space, but readership was lower.

It is only fair to point out that part of the low rating on sports items was due to the considerable disinterest of women; one should also note that the average male readership for sports items (22.1 per cent) was behind the average male readership for "Science-invention" (24.4 per cent). Something seems to have been garbled in transmission from researcher to publisher. A defect seems obvious, especially when we remember the traditional rationale for printing so much sports news in daily newspapers — "Giving the public what it wants" — is trotted out as the justification.

3. Sex tends to pull readers in different directions when it comes to science and health stories. Men were more interested in "Science-invention" (24.4 per cent as compared to 19.2 per cent for the female readers) while women favored "Health-safety" (22.3 per cent to 19.2 per cent).

The same general findings were uncovered in a study of the 3,353 photographs that appeared in the issues of the 130 papers surveyed. Percentages were the same with 1.1 per cent for "Health-safety" and 0.6 per cent for "Science-invention." Pictures grouped in "Science-invention" ranked ninth in "readership" among 35 categories, while "Health-safety" placed fourteenth. Photographs of fires and disasters ranked first; war views were second; and human interest pictures, including the "cheese cake" or bathing beauty variety, ranked fifth. Swanson's report suggests that women were more "picture-minded" than men.

From these admittedly ragged baselines (but the best we have), later studies and surveys have been made that show that the amount of space allotted to science, technology, and medicine has been expanding, and that with this has grown the public's appetite for even more.

The Associated Press made a qualitative and statistical analysis of its "A" or main trunk wire for six complete days in December, 1949, January and February, 1950. Results showed not quite 1,000 words each day for "Science," or 1.7 per cent of the total wordage; 1,350 words under "Aviation" were scattered over the six days for an additional 0.4 per cent. Thus, the combined total was approximately one science-aviation news item out of every 50 stories carried. This ratio tied in neatly with the "Continuing Study" figures for the previous decade.

A comparable study of the AP trunk wire copy and its Wisconsin state wire report for afternoon newspapers was made the following year, covering six dates from December 18, 1950, to March 3, 1951. For the national trunk wire, "Science and invention" totaled 1.8 per cent and "Safety and health," 0.5 per cent; on the Wisconsin wire, figures shifted somewhat with science decreasing to 1.0 per cent and "Safety-health" rising slightly to 0.7 per cent.

Prof. Scott Cutlip of the University of Wisconsin School of Journalism continued these studies for five selected days between December 15, 1952, and March 7, 1953. He found that the percentages for the trunk wire were 0.9 for "Science and invention" and 0.5 for "Safety and health." As in the previous study, the science news figure on the Wisconsin state wire fell to 0.06 per cent and the safety-health figure rose to 0.9.

In the 1949–50 study, well over a third of the "Science and invention" news on the state wire originated within the state (39.4 per cent), despite the relative decline in the portion of the full news budget. This meant that science stories by the press association's staff specialists in New York and Washington often gave way to items from within the state.

Conclusion: There may be considerable jockeying of news items at a press association relay point with stories written by science news specialists replaced by science items of more regional interest. Thus, a well-written science dispatch by a professional writer on a news service runs considerable risk of being "killed" at any relay point. Dissemination of science news, therefore, involves more than just making sure that it is properly written and then sending it on its way to a relay point.

The late Dr. Paul J. Deutschmann of Michigan State University surveyed New York City dailies for the first four weeks of March, 1959, and came up with what he called "mean stories per day" statistics. The average number of science stories in the seven papers then being published was 3.1, but they varied from 7.5 for *The New York Times* to 1.0 for the *New York Post*.

Prof. Guido H. Stempel III, then of Central Michigan University, used most of Deutschmann's techniques to survey eight Michigan dailies for the week of October 16–21, 1961. Four of Stempel's papers were fairly large ones — Detroit metropolitan publications were included — while the other four were smaller sized. Except for one of the smaller papers, percentages of news space for "Public health and welfare" exceeded those devoted to "Science and invention." Combined percentages of space given to both science and health news ranged from 9.7 for one of the smaller dailies to 4.0 for *The Detroit News*. The average figure for all eight papers was 5.7 per cent. Variations on allocations to science and health news seemed to have no real correlation with circulation size or whether the paper served a metropolitan area.

Dr. Edward M. Glick, then associate professor of journalism at American University in Washington, studied federal government-daily press relationships in the dissemination of health, education, and welfare news as printed in 22 large dailies in seven major cities during March, 1963. During that month, the Department of Health, Education, and Welfare issued 34 press releases totaling 16,000 words and, as a report in the *Columbia Journalism Review* (Spring, 1964) said, the major press associations "provided a steady flow of information based on both agency and Congressional sources." But the analysis found that 206 stories were published for a total of 1,992 column inches, with the press in the nation's capital alone providing a third of the published items and close to half the column inches printed. *The Washington Post* and *Washington Evening Star* printed a total of 73 stories, while only six were printed in Miami papers and five in Chicago. The West Coast cities of Los Angeles and Seattle published more than the two communities geographically nearer to the capital.

The *Review* summary said:

What it does mean is that the majority of the twenty-two large papers under study — including all the dailies in seven major cities — did not publish many of the stories made available to them by The Associated Press and United Press International. . . .

It seems obvious that the public in many metropolitan areas was not getting an adequate amount of significant information relating directly to its own welfare.

Conclusion: As with other topics, a paper's basic policies are a keenly determining factor in the decisions to print or to discard stories about science-related events and topics. This does not necessarily reflect an attempt to manage the news that gets to the public but simply the attitude that an editor's responsibility is to edit. But if the news executives have blind spots regarding science, the result is there for all of us to see.

In 1951, the National Association of Science Writers and New York University's Department of Journalism asked editors what was happening to their science coverage. While some subjectivity certainly entered the results, they are at least a rough approximation of the true picture. Almost two-thirds of the 50 editors who replied (62 per cent) said they had at least doubled the space allocation for science news in ten years. Eleven (22 per cent) said the increase had been approximately 50 per cent; four (8 per cent) reported slight increases, and two (4 per cent) said it remained the same as a decade earlier. None cited a decline during the ten years, which included the immediate post-atomic-bomb period.

Similar surveys on the first anniversary (1958) and eighth anniversary (1965) of the launching of the first Soviet Sputnik were conducted to try to ascertain the impact of that event and subsequent space shots on daily newspapers' allocations of news space to science, engineering, and medicine. Questionnaires were sent to every fourth managing editor on United States daily newspapers; out of the 400-plus news executives asked to participate, replies were received from 240 in 1958 and 166 in 1965. The results were:

	1958	1965
SPACE DOUBLED OR MORE	38.3%	47.0%
SPACE INCREASED BY APPROXIMATELY 50 PER CENT	36.7	30.1
INCREASED SLIGHTLY	17.5	18.7
NO CHANGE	4.6	3.0
NOT ANSWERED	2.9	1.2
	100.0%	100.0%

Not a single replying editor in either survey reported that his paper had reduced the amount of news space given to science items and, in both post-Sputnik surveys, more than three out of four editors had increased their allocation by half or more during the Space Age. Interestingly enough, the more recent statistics showed the heaviest concentration in the most extensive expansion.

One editor explained, "We do not cover 'science' because it is 'science' but because it is important news and as such competing with many categories of news."

Conclusion: The Space Age has brought a shifting of journalistic values in the news rooms of the country's dailies — just as it has changed assumptions in international politics and diplomacy. But one would be foolishly naive to assume this is due primarily to a suddenly awakened interest in "science for science's sake."

Science materials in magazines, like those in the daily press, have not received extensive research attention.

However, a content-analysis study of periodicals was made by Jerome Ellison and Franklin T. Cosser of the Department of Journalism, Indiana University; they dealt with non-fiction articles published in nine mass circulation magazines during the two three-month periods of September–November, 1947 and 1957. Under a system of allowing points if the article touched on several topics, "Science" ranked eighth for both periods out of ten subject matter categories. Science was also

one of six categories whose scores increased during the ten year
interval: approximately 4 per cent of the non-fiction magazine
articles published in 1947 and 5 per cent ten years later dealt
with some aspect of science. Despite the initial Sputnik launching that took place during the 1957 period studied, the authors
commented in *Journalism Quarterly* (Winter, 1959), "Even
in the age of satellites, editors are finding science hard to
sell."

Under a grant from the Magazine Publishers Association,
Dr. Robert Root of the School of Journalism, Syracuse University, studied all 3,763 articles printed in the 205 issues of
the ten more widely circulated magazines published in 1963.
The Audit Bureau of Circulation figures for these publications
totaled 72,873,649; thus they represented the cutting edge of
the mass circulation magazines' social impact that year.

Root found that 287 of the 3,763 articles (which included
fiction) dealt with science, which he defined as "space science,
natural science, medicine, engineering." This averaged 8 per
cent for the 1963 magazines or approximately double the Ellison and Cosser's 1947 figure. Incidentally, the science and
technology articles surpassed the 224 pieces of fiction that
Root counted.

The largest number of science articles — 89 or 12 per cent
of the publication's annual content — appeared in *Life;* 48
or 11 per cent were in *Reader's Digest,* 41 in *Good Housekeep-
ing,* and 39 in *Saturday Evening Post.* None of the other magazines printed more than 20 science articles during the year.
Three of the publications, all monthlies, printed less than 10
science articles; this accounted, in each case, for less than one
article out of every 20 published.

A study of mental health coverage in six mass circulation
magazines, which appeared in the *Journalism Quarterly* (Au-
tumn, 1962), indicated that at least one observer felt there
was a need for "competent, dedicated writers who will con-
scientiously refuse to 'write to order' for publications which

insist upon name-dropping and emphasis on sensational aspects of mental problems."

Conclusion: It is obvious that the intricacies of editorial policy complicate science reporting in magazines, much along the lines cited earlier for daily newspapers.

E. G. Sherburne, Jr., while director of Studies on the Public Understanding of Science for the American Association for the Advancement of Science, examined what science programs were available to San Francisco viewers who watched three television network affiliates, one independent commercial station, and one educational television network outlet during the month of March, 1963. He used a broad definition of science much along the lines of that given in the opening chapter of this book, because, as he explained, "Science is not a subject, but the warp and woof and frabric of our 20th century life."

During a typical week's prime-time hours of 7 to 11 P.M., a San Francisco viewer could watch 47 science or medical programs for 35 hours. Thus, 6 per cent of the total possible prime-time for the five stations was devoted to what is included under a broad definition of science.

Sherburne said that the significance of 6 per cent of total prime-time increases when we remember that these were the hours when masses of people watched; to illustrate, he estimated 42,000,000 across the nation viewed the weekly "Dr. Kildare" program with its by-product cargo of medical information.

Additional early morning hours and daytime programs for adults, as well as children's fare, added to the total science on San Francisco television. Although he did not make a detailed content-analysis, Sherburne pointed out in *Journalism Quarterly* (Summer, 1963):

In these daytime hours, as in the evening, there is more science than a precursory glance at the schedule indicates: a guest doctor, or a scientist (often a psychologist), may be interviewed on "Calendar," "Art Linkletter's Houseparty" or similar shows. In the early morning, there is "Continental Classroom," some news of science on "Today," and perhaps a bit of natural history for the younger child on "Captain Kangaroo." . . .

I would not venture a guess as to percentages across the board, but it probably would average out about the same as the prime time viewing hours. Especially it would be so if one took the viewpoint that the audience saw science even in the doctor-hospital soap operas.

Sherburne pointed out that while one city may at first seem a limited sample, the network operation of United States television, plus film and tape syndication, means that television offerings vary relatively little in American cities of similar size. In the study, offerings of the San Francisco educational television station amounted to 22 per cent of the whole five stations' allocation to science or only slightly more than the same proportion as the commercial stations programmed.

Conclusion: While the San Francisco study was admittedly quite limited, it may be more than coincidence that the great mass medium of television in the 1960's was allocating between 5 and 10 per cent of its content to items which qualify as science under its broadest definition. Thus, the findings for the broadcast medium would be in line with the earlier findings for newspapers and magazines.

Fashion trends and shifting interests often determine what gets into print and on the air. Again it is difficult to document; and, while there are some reports from individuals who have tested the news stream, this has been done even less often than the scattered information-gathering on content.

The 1951 survey by the NASW and NYU included a question asking editors to indicate those topics in science, technology, and medicine in which their dailies had special interest. Four out of five mentioned "Medicine and public health" and only a few less cited "Atomic energy." All the other topics were mentioned by less than a majority. I should point out that the question was asked more than a half decade before the first satellite launching.

The editors' replies squared with my own observations when I was a United Press science writer in Washington during the New Deal years. Stories from the U.S. Public Health Service and other medicine-oriented services almost without exception got better display than those reporting basic findings of research. About the only exceptions were stories that attempted the light touch, which, in retrospect, I fear must have pained — although I hope not really offended — some of the scientists who gave me the information.

When the NASW and NYU repeated a comparable study on the first anniversary of the original Sputnik, "Satellites and outer space" ran off with first place with more than four out of five editors marking that topic as having special interest for their papers. Slightly more than half of the 240 responding editors mentioned "Medicine and public health," which had been the leader seven years earlier, and "Atomic Energy."

This post-Sputnik newspaper enchantment with satellites and space was not temporary, as was shown by another survey on the eighth anniversary of the first Sputnik launching (1965). Out of 166 daily newspaper editors answering a survey question, 130 (78.3 per cent) said they had a special interest in "Satellites and outer space." Runners-up were "Medicine and public health" (69.3 per cent), "Agricultural science" (44.0 per cent), and "Atomic energy" (34.9 per cent). In the scant period of 15 years, this last category had dropped from second place in 1951 to fourth place in 1965. This drastic reduction in the ability of "Atomic energy" to compete for copy desk attention demonstrated the extent to which space ventures have replaced atomic energy in public interest — at least as seen from the viewpoint of newspaper executives. (See Table, p. 78.)

TABLE ONE

Areas of Editors' "Special Interests" in Science

(Based on NASW and NYU Surveys in 1951, 1958, 1965)

	1951	1958	1965
1.	MEDICINE AND PUBLIC HEALTH 82%	SATELLITES AND OUTER SPACE 80.0%	SATELLITES AND OUTER SPACE 78.3%
2.	ATOMIC ENERGY 76%	MEDICINE AND PUBLIC HEALTH 56.7%	MEDICINE AND PUBLIC HEALTH 69.3%
3.	AGRICULTURAL SCIENCE 40%	ATOMIC ENERGY 55.0%	AGRICULTURAL SCIENCE 44.0%
4.	NEW INVENTIONS FOR THE HOME 40%	AGRICULTURAL SCIENCE 32.5%	ATOMIC ENERGY 34.9%
5.	AVIATION 34%	MILITARY SCIENCE 28.3%	NEW INVENTIONS FOR THE HOME 19.3%
6.	RESEARCH GENERALLY 18%	AVIATION 27.1%	AVIATION 18.7%
7.	MILITARY SCIENCE 18%	RESEARCH GENERALLY 15.8%	INDUSTRIAL APPLICATIONS OF SCIENCE 18.1%
8.	INDUSTRIAL APPLICATIONS OF SCIENCE 16%	INDUSTRIAL APPLICATIONS OF SCIENCE 11.7%	RESEARCH GENERALLY 16.9%
9.	ASTRONOMY 14%	NEW INVENTIONS FOR THE HOME 10.8%	SOCIAL SCIENCES 15.7%
10.	SOCIAL SCIENCES 12%	ENGINEERING 7.9%	MILITARY SCIENCE 11.6%
11.	ENGINEERING 10%	ASTRONOMY 5.8%	ASTRONOMY 5.4%
12.	PHYSICS AND CHEMISTRY 6%	PHYSICS AND CHEMISTRY 4.2%	ENGINEERING 4.2%
13.		SOCIAL SCIENCES 2.9%	PHYSICS AND CHEMISTRY 1.8%
	N = 50	N = 240	N = 166

Pertinent to this shifting pattern in science news coverage was the comment on a questionnaire returned by Richard D. Smyser, managing editor of the *Oak Ridger* in Oak Ridge, Tenn., a town where the federal government has had scientific installations for more than a quarter of a century:

> In Oak Ridge, of course, we have a very special interest in science. It is our bread and butter. Generally, there surely continues to be a general upsurge in science interest and science reporting. I would note some tendency to report science in recent years simply because it is the "in thing" — the glamour and status subject. This has resulted in some superficial and poorly evaluated science news making some quite responsible columns. However, it would be my judgment that this phase is passing, if not past, and that science reporting is becoming more expert, accurate and interesting. Surely the period of some overplay was a small price to pay for finally digging science news out of the realm of the unknown and unwanted.

The Sherburne study of San Francisco television programming for 1963 showed that three-quarters of the prime time was devoted to medicine (59 per cent) and psychology (17 per cent). In other words, as the author reported, the television view of science is "predominantly one concerned with the ills and aches, the mending and fixings of man's sick body and mind." Despite the medium's preoccupation with entertainment, Sherburne found that 39 per cent of the medicine and psychology information broadcast to San Francisco viewers during March, 1963, was "interview, discussion, demonstration or lecture, or to put it succinctly, in 'fact' formats." These "fact" programs, along with the educational station broadcasting, showed the same bias in subject matter emphasis as did the other types of media.

Trying to discover why television had concentrated more on medicine and health, Sherburne wrote:

Medicine, I think, is popular for other reasons than the innate personal and human appeal. It is the practical science that has been with man the longest. The doctor, the practical artist of science, has lived among and been vitally associated with society for a longer period than any other of the practitioners and thinkers in science. Medicine is more completely incorporated into our thinking and general knowledge than other kinds of scientific endeavor. And its pragmatic approach is easier to comprehend. Further, the doctor by virtue of his professional role, is a communicator to the "common" man. These multiple considerations mean that the television writer, the producer, the advertiser and the viewer reflect a favor for the old familiar themes which we "feel" are easiest to understand and most significant personally. . . .

The first realistic conclusion one must accept is that the primary role of American television is one of entertainment. . . .

The challenge, properly put, is to make the entertainment, the everyday, every hour, ordinary run-of-the-mill programs more significantly involved and widely representative of the dramatic, exciting themes of today's world of science. . . .

The challenge is to our fonts of creativity — our writers, our producers, our actors — to see that new information, new ideas, new moral decisions and new conflicts surround the old themes of human struggle.

In regard to health programs on television, the President's Commission on Heart Disease, Cancer and Stroke in 1965 found that medium "ideally suited" for disseminating needed information to the general mass audience. However, the Commission lamented that health documentaries, to date, had been too few and some of those produced so poorly done that they were slotted into time periods unattractive to the viewer.

The Commission's Subcommittee on Communications reported to President Lyndon B. Johnson:

The health world has been slow to focus the awesome power of television on specific health problems requiring specific public understanding and response.

The medium is ideally suited for delivering clear visual information in dramatic and forceful terms. The art of the documentary film, true to science and at the same time challenging to the interest, is highly developed. Commercial television is capable of reaching an overwhelming majority of the American people, and educational television is growing rapidly.

Yet health documentaries have been few in number, uneven in quality, and generally drab in presentation. It has been their quality, rather than their subject matter that has relegated them to unattractive scheduling and doomed them to small audiences. Television producers are as aware as newspaper and magazine editors of the tremendous public interest in health. The products, with a few shining exceptions, have simply been inferior in the highly competitive world of commercial television.

Conclusion: Style and fashions often play large roles in editorial decisions, but subject matter also enters into final judgments.

6

THE SCIENCE WRITERS
Who They Are

Journalism has been called a "game" by some, a "profession" by others, and a "business" by still others. As a matter of fact, it combines elements of all three descriptions, but nowhere is its professionalism more apparent than among its "elite" reporters.

Ask almost any newsman who these elite groups are and he will be sure to mention Washington political writers and foreign correspondents. Unless he happens to be in the specialty himself, he will be less likely to include science writers — although a number of studies during the past quarter century have shown repeatedly that those who cover science rank with those who report national politics and international affairs.

The science journalist's professionalism is probably most obvious in his academic training. This should be no surprise since a major part of his assignments, like those of the Washington and foreign correspondents, involve taking a specialized technical language or environment and translating it, with a minimum loss of meaning, into the speech of the intelligent laymen unversed in the particular jargon and, hopefully, into the speech of the typical men in the street. This is no small task. For the science newsmen, it requires dual equipment: an understanding of science — its language, techniques, and philosophy — and a skill in communications that often has to transcend what is necessary to describe a dramatic forest fire or a courtroom clash of personalities. To prepare for this job means training in both prerequisites and, in an age of mass education, the traditional place to do this is in college.

Since the late 1930's, science writers have been surveyed in depth at least five times; Washington reporters, twice; and foreign correspondents, four times.* Thus we can construct

* See the following:

Hillier Krieghbaum, "The Background and Training of Science Writers," *Journalism Quarterly,* XVII, No. 1 (March, 1940), 15–18.

Lee Z. Johnson, "Status and Attitudes of Science Writers," *Journalism Quarterly,* XXXIV, No. 2 (Spring, 1957), 247–51.

Science, the Press and the Citizen: A Program for Public Understanding of Science (Mimeographed report of the Committee on Fellowships and Scholarships) National Association of Science Writers, 1957.

Report on Conference on the Role of Schools of Journalism in the Professional Training of Science Writers (Science Service, 1961).

William E. Small, *Training of the Science Writer* (Master's Thesis, School of Journalism, Michigan State University, 1964). Also summarized in "The Science Writer Survey," *NASW Newsletter,* XI, No. 4 (December, 1963), 11–13.

Leo C. Rosten, *The Washington Correspondents* (New York: Harcourt, Brace and Company, 1937).

William L. Rivers, "The correspondents after 25 years," *Columbia* a regular fever chart covering a number of points for a full quarter century, plus several cross-check points with other elite groups. Unfortunately, since phrasing of questions sometimes differed, it is like mixing thermometer readings in Fahrenheit, Centigrade, and Kelvin — without telling which is which.

On the basis of these studies, today's typical science reporter is a college graduate, probably in his forties and about 25 years out of school. Chances are strong that he took a concentration of science courses but probably did not major in any of them. His major, more likely, was English, general literature, or journalism.

Changes are overwhelming that he did not initially start his journalistic career as a science writer. In fact, unless he is a relative newcomer in the field, he did not really plan his college curriculum with an idea of covering science for the mass media. But science did not frighten him, either in high school or in college, and he never adopted the attitude of some liberal arts majors that the sciences could be neither interesting nor exciting. He probably took more courses in scattered fields of science than most of his classmates, except, of course, for those training to become research scientists, engineers, physicians, or teachers. Both physical and biological sciences attracted his attention and interest but he lacked extensive training in the

---

Journalism Review (Spring, 1962), I, No. 1, 4–10. Also see River's *The Opinionmakers* (Boston: Beacon Press, 1965).

Theodore Edward Kruglak, *The Foreign Correspondents: A Study of the Men and Women Reporting for the American Information Media in Western Europe* (Geneva: Librairie E. Droz, 1955).

J. William Maxwell, "U.S. Correspondents Abroad: A Study of Backgrounds," *Journalism Quarterly*, XXXIII, No. 3 (Summer, 1956), 346–48.

"The Foreign Correspondent: Survey Limns True Profile" (Study by Elmo Roper), *Editor and Publisher* (May 3, 1958), 14.

Frederick T. C. Yu and John Luter, "The foreign correspondent and his work," *Columbia Journalism Review*, III, No. 1 (Spring, 1964), 5–12.

social sciences — unless he is one of the younger members of the group. Chances are about even that he has done graduate work and, although they are far from typical, a few science writers have earned the Doctor of Philosophy degree and others have been awarded honorary degrees in recognition of their writing accomplishments.

He is a self-confident individual, in the main, and believes that he can write a competent news story or feature article for his newspaper or magazines, regardless of whether he has taken college courses in the areas discussed by the scientists. When he does have his doubts, the science writer is probably trying to deal with the abstractions of mathematics, engineering, metallurgy, or statistics.

Many of the "older hands" among the full-time science writers have attended a considerable number of "briefings" at which they received basic background or specialized information for interpreting new scientific developments. Like those elite scientists whom Dr. Derek J. de Solla Price claimed were commuting faculty members of the "invisible universities," the full-time, professional science writer also travels extensively. A majority of them will attend at least the annual convention of the American Medical Association and the post-Christmas sessions of the American Association for the Advancement of Science; some spend many days and sometimes weeks away from their home offices for missile launchings, professional society meetings, and inspection trips to laboratories and other facilities.

Financially, the typical science newsman is near the top of the journalistic pyramid, but even the most successful are far behind the syndicated columnists; there are no science writers who approach the national recognition and monetary rewards of Joseph Alsop, David Lawrence, Walter Lippmann, Drew Pearson, or James Reston.

There has never been a United States census of science writers and so, when anyone talks about this group, a fuzziness becomes apparent around the edges. However, a good guess would be that there are between 350 and 500, if one limits it to individuals in the mass media who spend half or more of their writing time on basic science, engineering, technology, and/or medicine.

The National Association of Science Writers, a professional group which has this half-time requirement for active membership plus at least two years of such experience, had 13 life members and 241 active members as of 1966, or a total of 254. Since life members are or were science writers, this means there are 254 newsmen scattered through all media, although representation from broadcasting and films is somewhat less than it is from print media. Among the newspaper reporters, the NASW membership represents virtually all those eligible, except the adamant non-joiners.

When Science Service in 1961 surveyed United States dailies and press associations, they indicated 477 special reporters handled science, in the broadest sense. But remember that this figure included those who handled two or three stories a month, as well as those who wrote nothing but science.

Science writers tend to cluster in those states that have extensive scientific facilities and numerous research personnel. But even more, they work in those states that provide the communication direction for the nation's mass media.

An analysis of the 1966 NASW membership list confirmed this distribution. New York City, sometimes christened the country's communications capital, was home base for 30.3 per cent of all the life and active members listed in the NASW. An additional 5.5 per cent lived in the metropolis' environs (including Northern New Jersey), so that one out of every three life or active members lived and worked in or near New York City. The rest of New York State provided only three members, one each in Albany, Buffalo, and Syracuse. The District of Columbia and its suburbs accounted for 12.6 per cent, while California listed 8.3 per cent, split slightly in favor of the San Francisco Bay area as against Los Angeles.

A breakdown of the 254 life and active members of the NASW by states showed:

	NUMBER	PERCENTAGE
1. METROPOLITAN NEW YORK	91	35.8
2. DISTRICT OF COLUMBIA	32	12.6
3. CALIFORNIA	21	8.3
4. ILLINOIS	17	6.7
5. MASSACHUSETTS	14	5.5
6. TEXAS	10	3.9
7. MICHIGAN	9	3.5

Much publicity in recent years has pointed out that a few states and, even more, a few institutions were getting the lion's share of research and development funds. The preceding table shows that the same concentration is true for science writers — at least for those who are life and active members in the NASW, which seems to be as good a measuring rod as we have. The three areas of New York City, Washington, D.C., and the San Francisco Bay region have half of the entire listing. Add the rest of California and the states of Illinois and Massachusetts — a total of four states and the District — and the percentage shoots past the two-thirds mark. Add Michigan and Texas and the total includes three-quarters of the group.

We could well ask whether the funds and facilities attracted the writers or whether the writers' activities helped to attract funds and personnel. Regardless of which came first, publicity for current projects has tended to keep additional grants and funds pouring in.

At the other end of the NASW listings, 24 states had not a single active or life member and 15 had no members at all. Associate membership goes chiefly to those concerned with public relations of science, technology, and medicine, or to those who do not write directly for mass media.

A 1964 analysis of 200-plus active and lifetime members

showed that approximately half were employed with newspapers and wire services. Among the remainder, 38 worked on
magazine staffs, approximately 40 classified themselves as free-
lance authors, and the rest listed miscellaneous writing assignments and teaching positions.

As might have been anticipated, the largest number of newspapermen who wrote science articles were with metropolitan
dailies. Statistics on circulation of their papers showed:

CIRCULATION 750,000 OR MORE DAILY	12
CIRCULATION 500,000 TO 750,000	12
CIRCULATION 250,000 TO 500,000	18
CIRCULATION 100,000 TO 250,000	39
CIRCULATION 100,000 OR LESS	11
NEWS SERVICES AND SYNDICATES	16

In connection with these figures, it should be kept in mind
that two-thirds of United States daily papers have circulations
of 25,000 or less. It is also obvious that papers with small staffs
will not assign one of their limited number of reporters to
spend a major portion of his time covering science, possibly not
even in a college town where education is the most important
local product.

Let us examine some of the details of these various studies.

In 1939, I surveyed the then-recognized 34 science writers
working on newspapers and wire services in the United States
(compiled from a list issued on June 14, 1939, by Science Service); because of my personal acquaintance with the group as a
fellow worker, I was able to obtain replies from all but three
of them, a return of 91 per cent. A comparable study of the
NASW membership was conducted during January and February, 1963, by William E. Small for a master's thesis at Michigan State University; he obtained 72 replies from active NASW
members, approximately a 40 per cent return. The figures show
what has happened to a reporting elite during the 24 years of
what has been described as the age of education.

	KRIEGHBAUM (1939)		SMALL (1963)	
	Number	Percentage	Number	Percentage
DID NOT ATTEND COLLEGE	4	12.9	4	5.6
ATTENDED COLLEGE BUT WAS NOT GRADUATED	4	12.9	10	13.9
GRADUATED FROM COLLEGE	23	74.2	55	76.4
DID GRADUATE STUDY	13	41.9	34	47.2
DID NOT ANSWER QUESTION	—	—	3	4.1
TOTAL NUMBER IN SAMPLE	(31)		(72)	

While changes were far from startling, there was a general upward educational mobility in all the categories; if the three who did not provide information had done so, the figures undoubtedly would have been even greater for higher education. Recently, the only people in science writing who did not attend college have been those who established themselves years ago and still continue to write. New recruits all have college backgrounds, and most of them have had graduate work.

Three other surveys, which did not give as much detail on educational background, confirmed the general trend:

• The 1955 Johnson study showed 79.0 per cent or 49 of the 62 science newsmen in his sample were college graduates.

• A 1957 NASW membership survey found 69.9 per cent of 176 reporters were college graduates; more than half of these (37.5 per cent) had also done graduate work.

• A 1961 Science Service survey of 249 newsmen, many of them part-time science writers, indicated that 81.9 per cent of the group had attended college. (No figures were given on those graduated.)

These statistics show a far greater emphasis on the academic than one would normally expect, even with the increasing trend towards higher education. Currently one college-age person out of three actually enrolls, and nearly half of those who go stay to obtain a degree.

How do science newsmen compare academically with Washington reporters and foreign correspondents?

Three years before my 1939 survey, Dr. Leo C. Rosten spent a year in Washington under a pre-doctoral field fellowship grant from the Social Science Research Council and gathered material for a book-length study of 154 correspondents there, 127 of whom filled out a detailed questionnaire. Twenty-five years later, Dr. William L. Rivers restudied the Washington press corps, using a larger sample of 273 political writers.

The 1936 study showed the science journalist well ahead of the political correspondent in college attendance (87.1 to 80.1 per cent); discrepancies became even greater as one tallied holders of bachelor and graduate degrees. Among the science writers, 74.2 per cent had bachelor degrees (as compared to 51.1 per cent of the Washington reporters) and 41.9 per cent had done at least some graduate study (as compared to 12.4 per cent of the political writers).

The 1960's River's study put the Washington writers abreast of the science specialists in college attendance and bachelor degrees, but the political correspondents remained well behind in the percentage that had done graduate work (47.2 per cent as compared to 31 per cent). During the quarter century between the two studies, college education had blossomed as a prerequisite for getting past many receptionists in personnel offices — including some at newspapers, magazines, press associations, and syndicates. Only among the younger political correspondents had advanced degree work become fashionable, whereas it had been fairly common among science writers in the 1930's and continued so in the 1960's.

A detailed study of foreign correspondents for United States publications was done by Dr. Theodore E. Kruglak during 1953–55, more than a decade and a half after the 1939 science writers' survey, the figures showed that the academic backgrounds of 130 United States nationals serving as full-time correspondents in Western Europe for American press associations, newspapers, magazines, and radio were below those of the science writers, especially in higher degrees and advanced studies. Those who attended college comprised 86.2 per cent of those Americans on overseas news assignment and 87.1 per cent of the science writers; figures for the foreign correspondents showed 11.5 per cent held advanced degrees and 16.9 per cent had done other work beyond their bachelor degrees; for science writers of half a generation earlier, comparable statistics were 41.9 per cent for both categories combined. We should also keep in mind that during that intervening decade and a half the science writers' education undoubtedly had been upgraded as better-prepared individuals entered this field. An improvement certainly is reflected in the Small study in 1963, almost a decade after Kruglak's. Then 47.2 per cent of the science writers reported they had done some graduate work.

It is noteworthy that three science reporters in the 1939 sample of 31 had the Doctor of Philosophy degree; in the only other survey to gather this information, one in 1955, two reported Ph.D. degrees, one an M.D., and one a Doctor of Science degree. Among the foreign correspondents in the middle 1950's, Kruglak found that three Americans held the equivalent of Ph.D. degrees.

During 1954 and 1955, Prof. J. William Maxwell investigated the backgrounds of 209 full-time correspondents, both United States and foreign nationals, willing to cooperate in such a study; they represented all the major American foreign news organizations around the globe, not just those in Europe as the Kruglak study had. College degree holders totaled 61.2 per cent; together with the non-graduate college men, they amounted to 90 per cent of the 209, thus ranking higher than the newsmen in Western Europe and the science writers of a decade and a half earlier. More than a tenth had earned master degrees (40 per cent of the science journalists had taken at least some graduate courses.) Two had earned Doctor of Philosophy degrees. Of the 127 who had received college degrees, 13 had won Phi Beta Kappa keys.

When Elmo Roper studied answers from 448 members of the Overseas Press Club of New York City in 1958, he found that 88 per cent had gone to college; 16 per cent had done graduate work; and 13 per cent had received advanced degrees, either master's or Ph.D.'s. (His report did not include figures on those who were college graduates.)

A few years later, Profs. Frederick T. C. Yu and John Luter of the Columbia University Graduate School of Journalism sought to survey all full-time American nationals reporting from overseas for major United States news media; they received 140 usable returns. Their results, printed in the *Columbia Journalism Review* (Spring, 1964) showed only 5 per cent had no college training and 57 per cent held at least one degree, with a third of these holding two or more degrees. This contrasted with Small's study where more than three-quarters of the science writers held college degrees and nearly half of the entire sample had done some graduate work.

Conclusion: During the past quarter century, science journalists have acquired better academic backgrounds and certainly have held their own against two other recognized elites among reporters — those in Washington and those overseas.

When it came to selecting a major subject for specialization in college, the science writers have, quite naturally, split their affection between topics related to writing and English and those related to the various sciences. The 1939 study showed that among 26 areas mentioned as majors or parts of double majors, exactly half were among the sciences; chemistry was the most popular with three mentions. English, general literature, and journalism, combined, were listed as majors nine times or 34.5 per cent. In the 1955 survey, the only other one to tabulate major undergraduate interests, 26 out of 62 respondents specialized in writing areas, while a minority picked science majors, with physics and chemistry being the most frequently mentioned, five times each. (The total number of science majors was not given.)

In comparison, Kruglak found that 56 of 130 foreign correspondents or 43.1 per cent had undergraduate majors in what he called the "social-political sciences," which included economics, history, government, political science, and international law and relations. The Maxwell study said one-fourth of the correspondents who attended college had majored in journalism — or "more than concentrated on any other subject." (Other majors were not listed.) Roper reported that 33 per cent of those who had gone to college had taken journalism courses.

Although a majority of science writers did not choose science for their major in college, the overwhelming majority did exhibit their bent toward the sciences by electing courses in a wide variety of these fields. The 1957 study of 176 NASW members showed that 82.4 per cent had taken one or more college science courses, with interest not quite evenly divided among the biological sciences (29.1 per cent), the physical sciences (22.8 per cent), and mathematics and engineering (21.9 per cent). As a sort of confirmation of an early intention to explore fairly deeply the college science curriculum, only 1.3 per cent reported they had taken an undergraduate "General Science" course — frequently the single requirement for non-science, liberal arts major trying to avoid making more than a gesture toward bridging the gap between the "two cultures." Social and behavioral sciences were largely neglected as college choices, with only 4.7 per cent listing sociology and anthropology and 2.6 per cent listing political science and history.

The 1961 Science Service report on a broader sampling of both full-time and part-time science reporters presented a different pattern of preferences. Most popular here were the physical sciences (59.8 per cent), followed by the social sciences (51.4 per cent) and the biological sciences (45.8 per cent). However, this analysis included 7.6 per cent who took medical science courses. No information was available on how many of those who took medical science had not, in addition, studied biological sciences. If the figure was substantial, it would tend to push the biological sciences up to a position not dissimilar to that in the earlier group.

The 1963 Small report also included information on the average number of semesters of college mathematics and science courses completed by science journalists. Far in the lead was mathematics (2.91 semesters on the average), followed by physics (1.87), chemistry (1.80), biology (1.29), and psychology (1.19). Here again the heaviest emphasis was on the physical sciences and scanty attention was paid to the social sciences.

The only study of parental occupations of science writers, the 1939 study, indicated that most were upper-middle-class occupations or professions. It appeared that a parent's economic status, which permitted him to send his child to college, was more important than any "inherent" or environmental interest in either sciences or journalism. In the sample of 31, seven parents were in business or manufacturing, six in science-related jobs (two chemists, a locomotive engineer, a mechanic, a pharmacist, and a physician), and five were in journalism-oriented positions (three newspapermen and two associated with the printing business). This emphasis on upper-middle-class backgrounds also was true in Rosten's survey of Washington correspondents; undoubtedly, similar findings would show up in any contemporary study of science journalists' parents, provided we allowed for a relatively easy entrance to college.

In 1939, the largest group (17 out of 31 respondents or 54.8 per cent) had done only journalistic work before becoming science writers and only one of these newsroom veterans had shifted over during his first year on a paper. The median year for being assigned to science reporting was the eighth, but one writer had 23 years of experience before he shifted and another had 20 years. The 1963 study showed that the average respondent had 14 years of professional science writing experience, backed up by 10 years of reporting experience. Six of the 31 writers surveyed in 1939 had done scientific research themselves, while 14 out of 72 reported they had done such work when questioned in 1963.

Conclusion: City newsroom experience or magazine staff work seem more impressive credentials than work in a research laboratory when editors and personnel managers are choosing people to write about science for the mass media. But rare indeed is the college graduate who lands a job in this field immediately out of school.

These same statistics on typical background also tell us that the average science writer, obviously well launched in a career, was, not unexpectedly, at least in his early 40's. A few professional science newsmen have continued into their 70's but, with pensions and retirement plans such as they are today, it appears unlikely that many of the current science journalists will be working much beyond their middle or late 60's in the years ahead.

Much the same pattern was found among Americans working overseas who were questioned by Yu and Luter. These foreign correspondents averaged 17 years of news work, of which 10 were spent overseas. They had been general-assignment reporters, rewrite men, Washington correspondents, and editorial writers.

In his 1963 survey, Small uncovered evidence suggesting that science writers are a generally confident lot who rarely feel that their stories suffer because they lack college-course training in the areas about which they are writing. The answers to two questions had provided this evidence: he asked the 72 respondents to appraise their own relative competence in science areas and to outline their own college course backgrounds. Here are the top areas in each group:

RELATIVE COMPETENCE	ACADEMIC TRAINING
1. MEDICINE	1. MATHEMATICS
2. BIOLOGY	2. PHYSICS
3. PSYCHOLOGY	3. CHEMISTRY
4. CHEMISTRY	4. BIOLOGY
5. ASTRONOMY	5. PSYCHOLOGY
6. SPACE TECHNOLOGY	6. MEDICINE
7. PHYSICS	7. GEOGRAPHY
8. GEOGRAPHY	8. ELECTRICAL ENGINEERING
9. ZOOLOGY	9. GEOLOGY
10. ANTHROPOLOGY	10. ZOOLOGY

While it could be misleading to generalize from such a small sample, one at least might speculate that the typical science writer feels at home in astronomy, space technology, and anthropology, despite his obvious lack of college course work. The National Aeronautics and Space Administration and other interested groups have held repeated briefings to fill in gaps in science writers' background so that, in effect, most of the professionals have had postgraduate, extracurricular training. In addition, space developments have been so rapid that college training of only a decade ago would provide little more than historical perspective; this would be especially true of technology and less appropriate for basic space sciences. Even there, "pure" science has shot forward as findings of space probes and satellite orbitings have been interpreted.

On the other hand, this comparison indicates that science writers, in spite of the college training that some of them have received, have some vague uncertainties when they report on mathematics and electrical engineering. According to the Small study, lowest competence ratings went (with "least competence" listed first) to home economics, civil engineering, mechanical engineering, metallurgy, statistics, agriculture, electrical engineering, and mathematics. The low ratings given to various engineering fields, mathematics, and statistics seem to show that reporters are well aware of the high degree of abstract or technical knowledge on which aspects of these fields rest. For the other sciences, they feel reasonably secure when they are assigned to write a story; this is true not only for the physical and biological sciences in which many had taken college courses but also for the social sciences in which many of the current journalists, especially the older ones, lack much academic training.

In fairness, I should point out that the full-time science reporter spends considerable time reading about new developments (frequently headlined as "advances"), and this includes not only the physical and biological sciences but also the medical and social sciences in which he may have taken few, if any, college courses. However, if science writers are like other adults who have been questioned, then they probably have carried some of the interest aroused by college courses into their "free-time" reading.

Like the more affluent professors and scientists who commute from conference to seminar to convention, science journalists in recent years also have been on the travel merry-go-round. Hundreds have gone to Cape Kennedy or the Houston Manned Spacecraft Center during satellite launchings and voyages — although newspapermen comment, sometimes bitterly and possibly unfairly, that the programming is geared to the overwhelming number of broadcast representatives. The print media still dominate the yearly science-convention features — the annual convention of the American Medical Association and the post-Christmas AAAS sessions. Both have the aspects of a sideshow with something for everybody who wants to report his own favorite topic and thus, both are magnets for the mass media reporters. The 1965 AMA convention in New York City set an all-time record of 577 individuals registered at the pressroom, with the largest segment (242) from professional and trade publications and 105 from newspapers and wire services. The 1966 AAAS meetings in Washington attracted 449 registrants in the pressroom, with 98 from newspapers and wire services, 211 from magazine, journal, and book publishing houses, and 55 from radio and television. During the 1960's, radio and television have discovered the newsworthiness of both these conventions with more and more broadcasters showing up as a consequence, but they still run far behind the print media in attendance.

During the past decade a widely accepted status symbol among science correspondents — and this is not to denegate the public information benefits — has been a trip to the Soviet Union, on which a series about Russian science could be written. Several dozen have gone to Moscow and beyond. Trips to European science centers by metropolitan newspaper specialists and magazine writers have become fairly common with jet transportation.

If the scientists who make news are going to travel to all

parts of the globe, then the science writers will not be far behind with their typewriters and their recorders. And, if science news is to be covered adequately and competently, that is undoubtedly the way it should be.

Like other reporting elites, science writers are well paid. When Small surveyed 72 NASW members in 1963, 15 per cent of them reported their incomes were above $20,000 a year. When Rivers surveyed 273 Washington correspondents about the same time, he found slightly more than 9 per cent were paid more than $20,000. The Columbia team questioned 140 foreign correspondents and found 34 per cent earned more than $15,000 (compared with 31 per cent for the same base in Small's study). Rivers also said the median salary of all Washington political correspondents in his sample was $11,579 or, as he pointed out, approximately $4,500 better than the national average for reporters, as determined in a 1960 study by the Associated Press Managing Editors.

Science writers certainly are keeping up with the Joneses among the reporting elites in an affluent society.

7

THE SCIENCE WRITERS
How They Work

Since each of the "worlds" that a newsman covers has its own
assumptions and philosophies, it is not surprising that reporting
techniques too shift as one moves from one area to another.
The goal always remains the same: Get the news. But how
the journalist does that is colored by the environment in which
those he contacts are accustomed to functioning.

He might gather information from a police inspector at the
scene of a slaying in a markedly different way than he would
find out about a proposed multi-million dollar corporate
merger. This change of pace is especially true for science
writers; their operations are bound in by the scientists' devo-

tion to presenting information to their colleagues first — to publishing in scholarly and professional journals or to giving papers before learned societies. The clash of these viewpoints was discussed in some detail in Chapter 3.

While some top-flight science reporters do go out foraging in laboratories and on campuses for news, most spend their time attending science and technical conventions, reading journals, and scanning press releases. More than in most other fields, such as politics, say, the news comes to the science writers. This does not mean that on occasions science journalists do not contend with the same pressures of news gathering and deadlines as their newsroom associates.

If a story is, in their opinion, of transcending importance as news, newsmen will concede little, if anything, to a scientist's request for a delay. The same approach prevails when the findings have entered the news domain through a scientist's presentation to his colleagues at sessions or in publications. But, for stories of lesser impact, the science journalist is generally willing to go along with a suggestion that the item be held until it is "ripe"; from the researcher's or inventor's standpoint, that might be until he presents it to his associates or until a patent is applied for. Every tidbit of research is not immediately played up in a column, Sunday feature, or on the air. It is a sort of "Live and let live" method of operation — except for "scoop" news. Thus, most of the time the science journalists work under slightly different conditions than many of their fellow staff members.

To social scientists, the science journalist functions as an observer, a teacher and guide, or an entertainer. To many working newsmen, this seems a bit far-fetched, but it may be worthwhile to take a closer look at these categories despite some discussion of them in Chapter 2.

As an observer, the science news specialist is translating and interpreting what he sees and hears about science so that his stories can be comprehended by the typical man in the street, the general public. This means dealing with specialized jargons, what might be called the dialects of the language of science, as well as with abstractions that often have no direct counterpart in the everyday world. Some scientists believe that this translating job can never be done completely. Every translation, whether from the jargon of a science or from the language of the ancient Egyptians, tends to lose some flavor in the process but it need not be condescendingly less intellectual.

As teacher and guide, the science writer may have to fill in the missing links in the backgrounds of his audience as he goes along because so much of what has been discovered in science, technology, and even medicine, has taken place since most adults completed their formal education. As Dr. Robert Oppenheimer once remarked, "Nearly everything that is now known was not in any book when most of us went to school; we cannot know it unless we have picked it up since." So, to give his story meaning, the newsman has to educate. And, to provide grist for the decision-making process by the general public, he also has to provide facts, some of which may be quite new and some quite strange to the non-science-oriented.

As an entertainer, the science journalist can operate like most of his colleagues in the newsroom and provide human-interest articles that will help to make scientists come through as human beings and not demi-gods — or devils. Or he may supply vicarious thrills — the dangers of space exploration, the shock impact that makes a reader say, "Gee whiz!" Or he may tell humorous stories, exhibiting the human side in that way.

In successful science reporting for the mass media, an essential, probably the very key, is to make the news understandable and intelligible to the laymen whom the writer is seeking to reach. This applies whether he is covering a convention, journal article, or press release. One of the ways to do this is to persuade the scientist to forget his jargon. The journalist seldom uses it and usage can easily get in the way of accurate reporting.

How this was done by one internationally known science writer, Ritchie Calder (now Lord Ritchie of Balmaslanner), 1960 Kalinga Prize-winner and professor of international relations, University of Edinburgh, was told in this reminscence at the first Inter-American Seminar on Science Journalism at Santiago, Chile, in 1962:

> Thirty years ago, as a young general reporter, who had not even begun to pretend to be a science-writer, I went to see Lord Rutherford, the father of nuclear science, at the Cavendish Laboratory, Cambridge. He had been the victim of journalists and he was not co-operative. I argued with him; I was trying to persuade him to help me to explain the nature of the nucleus to my 2,000,000 uninformed newspaper readers. He was scornful. He said they could not possibly understand. When I persisted, he opened the drawer of his desk, took out a manuscript of one of his latest papers. It was covered with hieroglyphs which I could not begin to understand but I pretended to study it, leaving him fretting for ten minutes at the other side of the desk. Then I took out my shorthand notebook, with the verbatim notes of a lecture which I had reported the night before. I threw it across the desk to him and said: "Lord Rutherford, I'll do a deal with you. You translate my shorthand and I'll translate yours. There is no more reason why I should understand your grammalogues than you should understand mine." Fortunately he had a sense of humour. He roared with laughter and settled down to give me a perfectly lucid, entirely comprehensible account of the nucleus which my 2,000,000 readers could understand. It was a great pity that Rutherford did not understand shorthand because that verbatim was an account of a lecture which George Bernard Shaw had given to The World League for Sexual Reform!

This translation and interpretation needs patience. Patience on the part of the scientist; patience on the part of the expositor and patience on the part of the reader, or the viewer, who has to be wooed into an understanding of the discoveries and developments which are dominat-

ing and revolutionising his life and the society in which
he lives. . . .

I came to terms with the scientists, thirty years ago.
The facts were inalienably theirs and any tampering with
these facts was reprehensible and a breach of our com-
pact. But the treatment and presentation were similarly
mine. My craft was that of the communicator; as a trained
journalist, I claimed to know what would attract, interest
and inform my readers and they had no more right to
tell me how to present my articles than I had to tell them
how to conduct their experiments.

Annual meetings or other conferences of the major scientific,
technological, or medical organizations are important sources
for news and thus meccas for science writers. A National Association of Science Writers' handbook explained why this was:

> To these conferences come leaders and active research-
> ers within a discipline. They present progress reports
> and results which may have significance to Society. There
> they discuss their particular problems, and present chal-
> lenges to the public for understanding and support.
> There they discuss policy. And there, increasingly, they
> meet and become mutually acquainted with science
> writers whose task it is to communicate to the public the
> advances and directions of the physical, biological and
> social sciences.
>
> It is the science writer's duty to present accurate and
> interesting reports of all these proceedings.

To a considerable degree, possibly more than they or the
scientists really realize, science journalists at a convention
decide what the news will be. They are the real "gatekeepers"
because if they do not write a story or shoot a film it is unlikely
that anyone else will — unless, of course, a maverick rival reporter at the same convention decides that he will not follow

the pack and comes up with a different talk or panel discussion in his featured lead or film.

Why is this true?

The available papers at such national conventions as the American Association for the Advancement of Science, American Chemical Society, American Institute of Biological Sciences, American Medical Association, American Physical Society, Federation of American Societies for Experimental Biology, and probably several dozen others surpass 100 a day. Thus a reporter could not possibly write about all the papers — even if he had the slightest desire to do so, which he does not. This forces him to select those half dozen or less that he will cover. He may do this selection by concentrating on those scientists who already have become popular figures, those that appear on television or in general circulation magazines, or those who hold high positions in educational administrations or political hierarchy. Or he may select those topics that he believes have a keen public interest. For example, a new pronouncement on the possibility of life on Mars would push out a new theory on how the computer might be better utilized to search the literature of atomic energy. And a panel discussion on how quasars are related to the origin of the universe would receive more journalistic fanfare than a new concept on the properties of special mathematical functions such as sine and cosine. These decisions are made *a priori* by the newsmen because they feel they can judge "what the public wants." Generally, they seem to be right — but one cannot avoid wondering how often they may be wrong. To what extent is the coverage of a scientific convention determined by journalistic stereotypes that may be obsolete because better educated readers, viewers, and listeners have joined the mass media audiences?

How press relations for a scientific convention operate have been described in a 1964 booklet, *Scientist, Meet the Press: A Practical Guide to Press Relations for Scientists and Research Managers,* prepared by the News Relations Section, Public

Relations Department, Smith Kline & French Laboratories. It said in part:

One of the busiest places at most scientific meetings is the press room. The press officer in charge greets reporters and science writers, briefs them on the program, takes care of their special needs, arranges press conferences and interviews, writes news releases, supervises the clerical staff, and spends a great deal of time on the telephone trying to bring scientist and newsmen together. The most important item in the press room (aside from the coffee pot, typewriters and telephones) is a file of the scientific papers presented at the meeting. All scientists who are to give papers should have some information on their report in that file, i.e., an abstract, science release or a copy of their paper.

The job of the press officer for a major scientific meeting begins many weeks or months before the meeting. For example, after reviewing abstracts on hand he'll contact some or all of the investigators on the program and ask for a copy of their papers. If one will not be available prior to the meeting, and the subject is considered to have news value, the author may be asked to furnish a news release (usually 50 copies) for distribution from the press room. Also, the experienced press officer knows that newsmen will automatically be attracted to panel discussions or papers delivered by scientists who have achieved previous recognition.

Newsmen covering a scientific meeting are alerted to story leads in several ways. The press officer may suggest certain research work considered significant or interesting. (Here he often consults a committee of advisors from the organization sponsoring the meeting.) Newsmen also scan abstracts on file or published in the official program, scan news releases available in the press room, and occasionally pick up ideas from conversations with other

writers. Once the decision is made to go ahead with a
story, the writer will try to get a copy of the paper. Also,
he may ask the press officer to arrange an interview with
the scientist. If more than one writer asks for an interview, a press conference may be set up.

There is another advantage of convention coverage, as explained by David Warren Burkett,* former science editor on
the *Houston Chronicle*:

> Any convention story can contain these three little
> words that unlock the heart of a city editor and conse-
> quently space in the newspaper. The words: "said here
> today." They reassure the editor that while he may not be
> publishing eternal truths he has a topic that is here, and
> now, and local.

Earl Ubell, science editor of the late *New York Herald
Tribune,* pointed out another advantage, long-range in this
case:

> One of the most important sources of experience is
> covering national scientific meetings. By covering such
> meetings, year after year, and reporting them to your
> newspaper, a science editor gets to know many scientists,
> their work and how they think. This is an invaluable
> storehouse of ideas for future stories and of awareness of
> stories which are to come.

How a not entirely typical press conference went just after
a delay in a satellite launching was described in the *Columbia
Journalism Review* (Summer, 1962) by Frank McGuire, then
a fellow in the Advanced Science Writing Program at the
Columbia Graduate School of Journalism who had covered

---

* A complete chapter in Burkett's *Writing Science News for the Mass
Media* (Houston: Gulf Publishing Company, 1967) deals with "Covering the Scientific Convention." See pp. 78–98.

missile launchings. It is significant because it shows the rough and tumble action that sometimes develops when reporters, scientists, engineers, and public information men get into a general free-for-all. McGuire tells what happened at Cape Canaveral (now Cape Kennedy) just after an additional launch delay had been announced for Astronaut Scott Carpenter's flight. The reason for the postponement, officials said, was a problem in the parachute-deployment system, which also had given Lieut. Col. John H. Glenn some difficulties.

McGuire continued:

> Mr. Bland, the technical expert, is introduced and begins to explain the electrical circuit alterations being made to insure parachute deployment at the proper time in the capsule's descent. He runs through the engineering modifications, describing barostats, redundant circuits, limit switches, interlocks and other details. Some of the technically adept newsmen ask why the change was made in this way, since it is obvious from the diagram that all the redundant circuits depend for their operation on a common barostat (altitude-sensing device) and if it fails, the entire system fails.
>
> Long explanations follow about pilot alternative action, statistical probabilities of failures, the function of the interlock. Not satisfied, someone wants to know when it was decided to make this change, since the postponement was announced shortly before the final countdown was to begin, and it looks like a rather hasty engineering "fix" instead of a solid solution.
>
> Mr. Bland begins to explain; Colonel [John A.] Powers interrupts with his own version. A voice from the press yells: "Let *him* explain it, Shorty, we know we're supposed to be stupid and we'd like to learn something." Dead silence. Colonel Powers continues, saying the decision to find the problem was made even before John Glenn's capsule touched the water, when he radioed about early parachute operation.

"Right then and there we decided to find the problem and make needed corrections," Colonel Powers says, "and if you've read your blue book [the NASA report of the Glenn flight, containing proposed capsule modifications] you'll see this is no hasty jury-rig."

A hand juts up from the press: "Shorty, I've got that blue book right here and there is no mention of NASA planning to modify the parachute system. Mr. Bland, when did you decide to change the system?"

Colonel Powers again begins to talk, but he is shouted down by several voices asking to hear from the technical expert himself. Colonel Powers continues: "Mr. Bland is an engineer, not a management-level executive." Mr. Bland looks relieved, and Shorty, executive-style, answers the questions.

Thus, consistently answering or attempting to answer questions put directly and pointedly to Mr. Bland, Shorty feeds the suspicion that Mercury project engineers made a potentially serious mistake, failed to find the problem, then rigged a series of multiple circuits designed to back each other up in case of failure. No satisfaction is gained by the press on the basic question. (Postscript: The modified parachute system failed to work properly, and Scott Carpenter had to use manual equipment.)

The provocative probe in press conference questioning is used by newsmen who feel that they are not learning the truth. When that happens, they will sharpen their questions in a persistent effort to cut through the subterfuge. A scientist, engineer, or physician who is frank and candid with reporters would almost never face the pointed cross-examination that confronted Colonel Powers.

His *modus operandi* pushes the science writer into extensive reading of specialized journals — scientific, technological, and medical, plus press releases from colleges, corporations, and government agencies.

At *The New York Times* science department each day, the

journals and press releases would comprise a pile between one and three feet high if they were opened and stacked flat. Victor Cohn of the *Minneapolis Morning Tribune* says he is lucky because he has to go through a pile only one to two feet high each week. Earl Ubell explained how the *New York Herald Tribune's* operation used to be:

At the New York "Herald Tribune" we receive some 100 technical journals from all over the world. Naturally it is impossible to read these. But we skim the table of contents of each journal and if something should interest us in that table we dip into the insides hoping to pull out a diamond of some kind. However, since the technical journals represent the mainstream of scientific information, we science editors get a good sampling of what is going on in the world of science. This experience is invaluable although I admit quite time-consuming. In short we do a lot of reading.

The question remains which of the hundreds — nay, thousands — of journals should a science writer receive? . . . Generally, in selecting journals one should divide the field of science into two parts — medicine and the physical sciences. By and large half of the science reporting which is done in the United States today concerns medicine — a very important and interesting field to the lay public. At the "Herald Tribune" we receive the following medical journals which we peruse rather carefully — the "New England Journal of Medicine," the "Journal of the American Medical Association," the "British Medical Journal," "Lancet," plus several other specialized medical journals. In the world of science in general, we get the "Scientific American," "Nature," "Science," the "American Scientist," "Biological Perspectives," to mention but a few of the many. All of these of course provide stories for our continuing coverage.

Representative of what the science writing professional considers to be a minimum of journal readings for those who want to keep abreast of current developments is a list of nine publications that are distributed by the Council for the Advancement of Science Writing, Inc., to its on-the-job trainees. This project, financed by the Carnegie Corporation of New York, is an effort to provide each participant with factual background for handling science news, some critical and self-confident judgment in the field, and knowledge of the special problems of science writing and how others have solved them. Its primary goal is not pointed toward training a full-time science reporter but rather toward helping those who on many non-metropolitan newspapers are asked to cover science stories when such occur in their home towns. The current list of journals supplied to on-the-job trainees includes:

> *Chemical and Engineering News*
> *International Science and Technology*
> *Journal of the American Medical Association*
> *Medical Tribune*
> *Medical World News*
> *Science*
> *Scientific American*
> *Today's Health*
> *Understanding*

Four of these publications are concerned almost exclusively with medicine and one, *Understanding*, is published by the American Association for the Advancement of Science and CASW "as a contribution to the growing efforts to bring science and the public closer together." The other four cover broad segments of contemporary scientific activities with more emphasis on basic research than on technology. However, the applications of science could not be avoided if a science writer diligently read this minimum list of journals.

After conceding that one reward of science reporting was "to uncover on one's own a good science story from an original journal report," Nathan S. Haseltine, medical writer for *The Washington Post*, suggested that probably some specialists spent too much time trying to read "a plethora of science journals." Then he added in the *NASW Newsletter* (March, 1966):

> This is not to say that science writers have no need for the journals or the scientific meetings. They are the pegs on which we hang most of our stories. But, unless our friends tell us what to look for, our output would be considerable less both in volume and quality than it is now.
>
> We need the scientific journals, not as prime sources of news but to check back on the material we are told is being, or will be, reported. To fail to make such checks would be to turn our responsibility over to others. After all, they, our news sources, have products or reputations to sell. Only by checking can we keep from over-selling them.

As mentioned earlier, established science reporters are flooded with press releases in addition to scientific journals. Some of these handouts, as they are called, are invaluable tips and background for news items; others are efforts to obtain unpaid advertising in the news columns or on the air for a new product or for a new idea in which some company, group, or institution has a special vested interest. The sophisticated newsman can spot both approaches with little difficulty, but the volume of mail, at least, has to be opened and scanned for news possibilities.

In the newsroom of one New York City daily, a "duty" science writer, to borrow a term from the armed services, gets this assignment and it takes him most of his working hours to go through "this haystack of material." A time-consuming job to be sure, but out of this "haystack" come ideas that may blossom into front-page science stories, magazine articles, or broadcast program ideas.

Victor Cohn of the *Minneapolis Morning Tribune* complained that public relations men of over-eager companies, governmental agencies, and universities could become "the guilty mad mimeographers of science and technology." But he added that many, on the other hand, did "an excellent and restrained job" and then, he said, "I try to read them."

One metropolitan science editor claimed that 99.9 per cent of the 50 or more press releases that came to his desk each day were "without value for newspaper stories," but that they did give him a kind of running commentary on what was going on in the industrial, governmental, and university fields, "when he can take the time to read them assiduously."

When the staff of *Understanding* surveyed NASW members in 1965, the 80 respondents reported that they received from 20 to 300 press releases each week. The median number of releases was approximately 100. The typical respondent either used immediately or filed about 7 per cent of the releases that he received; the rest were discarded.

When asked their preference on how they should obtain important news, two-thirds of the "fast media" respondents (those with newspapers, wire services, and news magazines as against those with other magazines with tardier deadlines, free-lance writers, and book editors) said they favored the press conference over the news release. As one wrote on his questionnaire, "If story really important, writers will want to ask questions, which gets you into a press conference."

The *Understanding* survey also asked, "Which institutions do the best job with news releases?" The 80 responses were summed up in the Fall, 1965, issue of the magazine.

> The most votes went to Harvard, Stanford, Chicago, Berkeley and Wisconsin among the colleges; to the space agency among government agencies; to the American Medical Association and American Institute of Physics among associations; to General Electric in industry; and to Roswell Park Memorial Institute [Buffalo, N.Y.].

Howard Simons of *The Washington Post* and twice winner of $1,000 awards from the AAAS and Westinghouse Educational Foundation once explained his activities on the "science beat" in the nation's capital, which has a wealth of agencies, bureaus, and offices:

> The best I can do is get around to each of the agencies maybe once every three months or once every six months. What is a scoop? The fact that I'm the only guy who has time to pay attention to a particular project at an agency? You can't ignore competition, but to be competitive at the cost of distortion is a sin. If a scoop is an informed story before an artificial release date set by the agency that's good, but rushing into print just for the sake of being first without accurate information is bad.

Carl Heintze of the *San Jose* (Calif.) *Mercury* and *News,* which have a combined daily circulation of well beyond 100,-000, described himself as "a science writer of sorts" and his reporting work "in the hinterlands" as "reducing science and the scientific revolution to their lower common denominators, explaining the significance of DNA experiments to those more interested in the Giants' standing in the National League race, sorting out the confusion of the thermo-nuclear bomb to those whose greater problem is disposal of an increasing supply of polluted air and sewage, two other important man-made evils."
He explained in the *NASW Newsletter* (December, 1963):

> Science to our editors means different things. It means you talk to the city health department about a diabetes screening clinic; it means you do the handout rewrites for the National Foundation and the Cancer Society, and that you are a kind of amateur doctor who diagnoses the ills of the staff and their relatives.
>
> It means you supposedly know something about local medical politics; the confusion which spreads around county hospital administration, and the three new hospitals which have suddenly sprung up in a mushrooming population like the subdivisions around them in uncertain provisions for the future.

It means . . . coping with space and technology, for both are important words in California, not because of science, but because of business dollars, payrolls and the harder realities of getting to the moon. It is a sort of ground level worm's eye look at the lofty view which many other NASW members see.

Just as any specializing reporter for print or electronic media follows the entire news picture and adds any appropriate contribution, so the science journalist types out or films stories on the scientific or medical angles of such events as, for instance, congressional appropriations for new research facilities; the illnesses of famous athletes, widely-known politicians, or popular entertainers; and such acts of Nature as earthquakes and hurricanes. This piggy-back coverage through a "with" story is old-hat procedure in newsrooms and it is not unusual that science writers operate much as do other specialists.

One writer called this the first "principle of usefulness" in guiding science journalists outside a restrictive news category. And, as he pointed out, it helped them earn their keep in good will and promotions from city editors and other news executives.

Some news specialists tend to have a highly vested interest — almost a favorable bias — in the areas they report. Rightly or wrongly, this charge has been leveled at United Nations correspondents.

Often there may be just the opposite reaction. To illustrate: David Warren Burkett, former science editor of the *Houston Chronicle*, was often involved with Houston's Manned Spacecraft Center, a major part of the "science beat," and related with unconcealed pleasure how he and other correspondents got around road blocks to their efforts to obtain fuller coverage of the NASA efforts. In his book, *Writing Science News for the Mass Media* (1965), Burkett gives a whole section to "commando tactics" which will "produce accurate, newsworthy stories without the approval or even cooperation of government administrative and information officers."

Some science correspondents also wonder if they are too impressed by the status commonly associated with widely-known scientists, engineers, and physicians. To illustrate, Henry W. Pierce of the *Pittsburgh Post-Gazette* wrote in the *NASW Newsletter* (March, 1966) that science journalists may be "a bunch of patsies" because many of them "have been suckered into an uncritical acceptance of anything we are told by our authorities — our authorities being doctors and scientists." He said the political writers, the police reporters, and the financial editors, all had a healthy skepticism toward most of their news sources and then added:

> But we, bless us, go in with our bright baby-blue eyes all aglitter, our pink little tongues dripping with eagerness, and, pencils poised, faithfully record anything our scientist-gods tell us. Never does it occur to us that these guys, too, may have motives that are less than noble.
>
> I do not mean the scientist who is out to make a buck or the man who likes to see his name in print. I am referring, among other things, to the guardians of scientific dogma, the men who reject new ideas simply and solely because they go against accepted scientific theories. It should be our job to look skeptically at the whole field of science. . . .
>
> Few laymen are aware of the kind of chicanery that goes on in the name of research at some universities — the deliberate attempts to fog reports with "scientific" verbiage, the plagiarism, the needless duplication of oft-repeated experiments. But what are we doing about it? We eagerly trot up to our sacred authorities, mouths agape and notebooks open, in pursuit of our other god, The Story.

A professional science writer's sense of professionalism drives him to insure, to the best of his ability, that the facts and implications of his copy and film are accurate and true. As one well-known correspondent explained, he "checked and double-checked five times" in his efforts to insure accuracy. It is not enough just to quote accurately — especially if the source is unclear or, even worse, unreliable. Every science writer worth his pay check has his own "little list" of experts that he respects, that he knows fairly well, and whom he feels free to call any time to check some point that bothers him. As Earl Ubell once put it: "Every science writer I know who thinks seriously of his job does have an advisory committee in his little black book (along with other addresses)." A writer may ask, "Is Doctor X's work really a significant advance? Or is it just an attempt to get publicity so he can obtain another grant?" Since the contact knows that he will not be quoted — at least, not by name without permission — and since he knows that he can trust the reporter, the scientist or physician answers candidly and frankly. He also performs a most invaluable function in the popularization of science: monitoring the factual accuracy with an expert eye for the individual who is going to write up the news.

But to have scientists or physicians appoint contact officers who formally "clear" news stories before publication infuriates newsmen because, as they repeatedly have said, it insults them and their abilities. Most of the regular science writers feel that it infringes on their professionalism and, as one said, "makes an *a priori* assumption of incompetence." Scientists and doctors protest their innocence of such implications but their statements just do not ring true to the reporters.

One well-informed newspaper editor who has had some of the country's best science writers on his staff suggested, tongue in cheek, that he looked forward to the day when a journalist's code allowed him to put on a white coat, go into an operating room, and pick up the scalpel if he did not like the way things were going at a hospital. He might have added that the complainers almost never visit the journalist's home newsroom and seldom the convention pressrooms.

Newsmen repeatedly have insisted that such errors as they make are seldom intentional. For instance, Arthur J. Snider of the *Chicago Daily News* told a University of Pennsylvania symposium:

> The fact is that there are a lot of mistakes in science writing. Some of them come from lack of information. They are not errors of intent. Only rarely today will an individual deliberately flog up a sensation to get a scoop.

And Ubell put it this way:

> Now, if any science writer has any inclination or indication that he is not up to transmitting information to the public, or that he thinks he has it inaccurate, he will call back and he will want help. I don't know of a single one of my colleagues who cynically and consciously wants to be inaccurate. And that goes for every newspaperman I know, by and large. He may be lazy, there may be other things involved, but not consciously does he want to be inaccurate; so he will try to get authentic information.

The question of "authentic information" presents a prickly problem. What Dr. A considers a major development may appear to Dr. B as an unsupported finding of a not-too-careful research worker. Unfortunately for science newsmen, there is no uncontaminated fountain of ultimate truth to which they may turn.

In recent years, there have been just enough times when a consensus of so-called "experts" turned out to be unequivocally wrong — after additional information was in. When a scientific group assured the United States government that a high-altitude shot to produce an artificial radiation belt would have no long-range damage, all seemed well. But after the shot-firing, one of the scientists confessed, "Well, I guess we were wrong." When a Polish refugee suggested during the 1950's at a session on cancer research that his work suggested viruses might be the cause of leukemia in mice that were normally resistant to the disease, many of his fellow scientists tried unsuccessfully to dissuade the reporters from writing up the refugee's remarks. Yet, in 1962, that same researcher won a $10,000 United Nations award for his discovery.

Even double-checking with his contacts may present booby-traps for a science reporter some times as Robert D. Potter, charter member of the NASW and more recently executive director of the Medical Society of the County of New York, explained:

> As a good reporter, the science writer will want to take what is said and report that, but he will probably want to check the statements with his own sources of information, and it is here that a weakness may appear. His best contacts are usually with the true research scientist and these are usually full-time professors of medicine who are on salary and who devote their lives to research. Some of these men of medicine from the research area may be just as naive on socio-economic questions as they are learned on the scientific side of medicine, and their reputation in one field cannot always be translated into authoritative statements on the other. The walls which separate the "town and gown" of medicine are the walls of the research tower compared to the general practitioner of medicine down the street. These are not mere physical walls of research institutions alone, but often a differing philosophy of how medical care shall be rendered and how it shall be paid for.

Undoubtedly easy access to the truth would make science writers' lives easier, but there is that age-old query, "What is truth?"

Science writers, at least those in such organizations as the National Association of Science Writers, have their ethical standards. While they are recent in adoption and have not had

to stand the erosions of time, they do seem to have considerable force.

In June, 1960, the NASW adopted the following as the first
step toward a set of ethics for science writers:

> 1. A science writer shall take all necessary measures to
> insure that the information he purveys to the public is
> accurate, truthful and impartial.
>
> 2. A science writer should not for any remuneration
> by a commercial organization permit his name to be used
> to promote a commercial service, a commercial product
> or a commercial organization. Such activity shall be con-
> sidered prejudicial to the best interest of this association.

As newsmen admitted, such questionable activities were
never common but they did exist.

Alton Blakeslee of the Associated Press and one of the most
widely read science writers in the United States reported at
the University of Pennsylvania symposium on communications
and medical research of an experience that he had had. He
said:

> Recently I was approached by a man who said he had
> an opportunity for me to place an article in a magazine
> on a free-lance basis. He described very frankly his own
> rather curious organization. He and his associates were
> representing a company which had developed a new
> product to treat a very common ailment. They guaranteed
> to find the medical researchers who would test it, and had
> done so. Further, they had a method of getting it pub-
> lished more quickly in a medical journal than might
> otherwise be done, so that it became "legitimate" news.
>
> At this point he went to a magazine and suggested a
> story on the general topic, and told the magazine editor
> that the company would place a large amount of ad-
> vertising with them if the story were used. He also volun-

teered to find a science writer who would write the story, and this is what he was talking to me about. He said I would make my deal with the magazine editor, and perhaps be paid $1,500 or $2,000 for the article, and all I had to do was to mention this new product by trade name twice, and never mention any other product. The company, he said, knew that writers were never paid what they were worth, so the company would give me $5,000 on the side. Then if the article were picked up and reprinted by a certain outlet, I would get that reprint fee, and the company would be so delighted with the advertising achieved that way they would pay me $10,000 more.

This added up to $17,000 — as I mentally kept track of the arithmetic — but apparently something else showed because when he had finished this explanation, he asked, "Why are you looking at me like I just crawled out from under a rock?" And when I tried to explain to him, as gently and as easily as I could, why I didn't want to have any part in such a planted story I simply said that I'd like to be able to sleep at night. He had a curious answer. He said, "I wish I could."

Not to belabor the point but to make it crystal clear, it was a professional science writer who declined to go along with the proposal after medical researchers and journal editors were perfectly willing to perform their assigned roles.

Nathan S. Haseltine of *The Washington Post* cited a nuts and bolts factor that tended to keep the science writers basically honest when he said:

I know of no other field in newspaper writing where dishonesty is so quickly recognized, and deplored. The science writer who plugs products, or friends, to the point of dishonesty, quickly loses esteem — and the successful science writer builds his success on the esteem of the scientists.

What about the broadcast medium's operations?

Only a few of the printed media reporters have tried their typewriters for radio and television. One of the most successful in this small group that have comparative experiences for these not entirely similar communications is Mildred D. Spencer, writer with the *Buffalo Evening News,* 1965–66 NASW president, and 1965 winner of an American Medical Association writing award. She described her work in the two fields in the *NASW Newsletter* (December, 1961) as follows:

> Producing a medical show and writing a medical story use two entirely different sets of mental muscles. The calisthenics of switching back and forth helps keep me mentally limber. . . .
>
> A successful medical program has to please two audiences — the lay people who watch it or listen to it and the physicians who seldom do either but are made acutely aware of it by their patients.
>
> The lay audience is the easiest to please. It prefers programs dealing with common problems in health and disease — backache, tiredness, arthritis, overweight, indigestion, heart trouble, cancer. It wants to know about new developments in the clinical field (for example, how to evaluate the relative merits of live vs. killed polio virus vaccines), but has little interest in highly technical research discoveries. Although it is vaguely aware that all physicians do not agree on methods of diagnosis and treatment, open disagreement among panelists makes it uneasy.
>
> The real challenge lies in keeping the goodwill of the medical profession itself. So far the Round Table has done so. Nearly 300 physicians have appeared as moderators or panelists. None has ever refused because he disapproved of the program's aims or the way they are carried out. . . .
>
> The program doesn't aim at drama and excitement. It tries to be informative and helpful, to alert the public

to the danger signals of disease without arousing fear, to discourage self-diagnosis and encourage medical checkups.

To avoid charges of "advertising" or "favoritism," Miss Spencer said that no physician was asked to appear more than once a year and that participants were chosen from all sections of the city, medical specialities, ethnic and religious groups.

In some communities, physicians in private practice are especially nervous about being subjected to the charge of possible "advertising" when they appear on radio or television or have their activities reported in the newspapers or magazines. Those in a few localities may have to explain their activities to the local medical society's board of censors. However, doctors who have gotten such publicity on the air and in print, agree almost without exception, that they got few, if any, good patients this way. As one exclaimed, those affected by it usually are "crackpots" or "deadbeats."

Miss Spencer said that for the Buffalo program she chose the moderators, selected participants from lists given to her by each moderator, suggested topics, and prepared statistics and other general information that the panel leader might not have at his finger tips. Then she added that she avoided "at all costs any intimation that I, a layman, would presume to run a medical show."

One scientist who turned television producer had this wistful comment on a not-infrequent occurrence when scientific colleagues faced the camera for the first (or even later) time:

Unfortunately, come air time, he spent most of that time staring fixedly into the camera, overcome by a sudden attack of silence.

But, he added, this was not always the case:

Those whom I and many of my colleagues have always considered to be the "greats" in their field were able to speak in simple English terms — demonstrating even the most abstract concepts with hand gestures or clear diagrams on the blackboard. The ones who took refuge behind the jargon-bush were those who might generously be called "developing," or honestly called "mediocre."

E. G. Sherburne, Jr., who directed public understanding of science studies for the American Association for the Advancement of Science until he became head of Science Service in 1966 and who previously worked in educational television on both East and West coasts, said that television's coverage of the Gemini manned space flights put this aspect of science and engineering "in the same league as such video spectacles as political conventions or the World's Series."

Sherburne assessed the pluses and minuses of such coverage in the *NASW Newsletter* (March, 1966) as follows:

TV's greatest powers lie at opposite ends of the reality scale. At one end it takes people into a world of imagination and fantasy through dramatic presentation and, at the other, it gives them a closeup view of great events in the real world as they happen.

The coverage of the manned space flights has been at its best showing the real-life drama of the missions — suiting up, entering the spacecraft, and liftoff, and interviews with the control room staff and astronauts. . . .

Specialists are rare in the TV world. As a result, general assignment reporters have had to be used in a number of situations. Their problems have ranged from inability to use simple technical terms to errors of fact to ignorance of equipment they were demonstrating. . . .

Another problem has been that of production quality. Too often the TV picture merely shows a reporter talking, occasionally handling a piece of equipment or a model, or referring to a map. NBC is the only network to use the television picture with any degree of creativity and its coverage has a dimension that is lacking in the coverage of its competitors. Such visualization costs money, and it is interesting to note that NBC is the only network with a sponsor for its space coverage. NBC's production quality may be due to the fact that it has a sponsor, or it might be the reason that NBC was able to get a sponsor.

What of the future? We can expect the reality coverage of the space effort to become more and more spectacular. Today we have countdown, liftoff, and splashdown. Tomorrow we will see an increasing amount of various kinds of film — rendezvous, extra-vehicular activity, docking, and transfer between spacecrafts. Ultimately, we will have live television from the spacecraft itself and from the surface of the moon.

An important characteristic of reality coverage is that it is commonly accessible to all three networks, either because it comes from "pool" facilities or because it is on film. This means that this is an area in which the networks cannot compete.

If the networks want to have their own space race, it will probably be in the area of supplementary coverage — commentary and interpretative programming. It would also bring them into more direct competition with newspapers, whose own response to television has tended to more interpretative material.

8

THE SCIENCE WRITERS
Some Work Histories

The behind-the-scenes details of how science reporters work are not often revealed but, when they are, they provide illuminating glimpses on how these news specialists go about their routines, sometimes with a touch of genius that gives them praise, promotions, and eventually profits.

To gather the news, science writers sometimes have to break through artificial embargoes to release the information they want to dig out, as illustrated in Case Numbers One and Two. Other times they have to battle unfavorable conditions of Nature, such as those in Antarctica and the Galapagos Islands, shown in Case Numbers Three and Four. Still other times it is a running battle to beat or, if need be, live with, administrative procedures set up by a government agency, shown in Case Number Five. In still other assignments, it is just a relentless quest for news from whatever source the reporters can tap, as in the example of President Lyndon B. Johnson's gall bladder operation in 1965. Case Number Seven demonstrates that broadcast science newsmen have their own journalistic fun and games. In all the cases cited, the efforts produced stories or broadcasts that were distributed and praised around the country.

Here are some work histories that are not atypical of what other professionals might have done, if they had had the opportunities.

CASE NUMBER ONE

Dr. Howard A. Rusk, New York University professor of physical medicine and rehabilitation and associate editor of *The New York Times* who writes a Sunday medical column, outlined the background on one of the greatest medical "scoops" during a University of Pennsylvania national symposium in October, 1963. His recital disclosed that a professional science journalist can employ ingenuity and talents to as great a degree as any other newsman. Rusk explained how William L. Laurence obtained an exclusive story on cortisone treatment for arthritis. Laurence has two Pulitzer Prizes to certify his status as a journalist, one for covering the birth of the Atomic Age and the other shared with four other reporters at the Harvard Tercentenary celebrations of 1936.

Here is the story as Rusk told it:

> At the American College of Physicians meeting in Boston in the spring of 1949 it was a well-advertised secret that at the Mayo Clinic there had been discovered a substance. As a part of the secret everybody knew it had to do with the adrenals, that it had almost magic effects in the immediate alleviation of symptoms of rheumatoid arthritis, that it had to be taken regularly, like insulin. These facts were well-known.

It was the hope of the Mayo group that had done the basic research to withhold the announcement of the discovery until the International Rheumatism meeting in the United States the following July.

Well-known secrets with the possibility of a scoop do not go unrecognized or unfollowed in the medical profession, and through a series of circumstances it was learned by one of the editors of *The New York Times* that the first announcement and the movies of this discovery would be shown at a regular Mayo staff meeting on a specific day in the early spring, that this was an open meeting and anybody could go.

With this knowledge, Mr. William Laurence, who was at a meeting in Detroit, was called. He was at dinner at the time. He had an hour to catch his plane, the only one that could get him there. He never returned to the dinner table. He never packed a bag. He got on the plane and went to Rochester. He sat in the meeting and made notes, much to the irritation of the investigators, about this new discovery. He heard the reports and saw the film. He had an open line to New York and he called the story in. Bill says that next to the atomic energy story it was the greatest scoop of his career. . . .

I tell the story because it shows that when you have a well-advertised secret about an impending great new advance or discovery you cannot hide it. You better announce it when you can within the framework of totality rather than to try to save it for a specific event.

CASE NUMBER TWO

Five days before the United States Public Health Service released its historic report on the relationship between cigarette smoking and cancer, John Troan, then Scripps-Howard science writer in Washington, D.C., sought to pick up the details so that he could beat his fellow writers. On a Monday, January 6, 1964, the Public Health Service announced that the report, long awaited by both physicians and the general public, would be released the following Saturday. Troan started by calling some of the co-workers of committee members since he knew none of the top officials well.

What happened is told as follows by *Understanding,* quarterly publication of the American Association for the Advancement of Science and the Council for the Advancement of Science Writing, Inc., in the Spring, 1964, issue:

> A few phone calls to these acquaintances turned up nothing. So John appealed to their egos. "But, Bill, you don't mean to say that (committee member) hasn't told even his *friends* about the report, do you?" This technique opened buttoned lips.
>
> Now Troan checked with three people who had seen the report, asking them, "Of course you won't tell me what's in the report, but at least tell me what's in my story which is way off base." This they did, and John had his two-day jump.
>
> There was nothing anyone could do about that scoop, but PHS did have a neat procedure for breaking the news:
>
> At 9:45 a.m. on the 11th, newsmen were locked in the State Department auditorium. Nearly two hours' time was provided for skimming the 387-page report and for a 30-minute press conference. Then the 250 reporters were freed to phone in their stories.
>
> The locked doors prevented itchy reporters from dashing to the phone with "quick flash" stories.
>
> The same idea was used by the Securities and Exchange Commission last year, in releasing its report on alleged rigging of stocks.

CASE NUMBER THREE

The South Pole and the United States' efforts at scientific studies in Antarctica have had such fascination for science writers that most of those who have been in the field for a decade or more have made at least one trip there. The United States Navy and the National Science Foundation, both of which have vested scientific interests in the areas around the

South Pole, have cooperated in trying to make the correspondents' visits as hospitable as possible.

Blair Justice, former science reporter with the *Forth Worth* (Texas) *Star-Telegram* and now with *The Houston Post,* told about his experiences in an article in *The Quill* (February, 1961) as follows:

It started like this: At 18,000 feet over polar waters, I reached under my sea bag for the portable typewriter accompanying me on a five-week Antarctic assignment. With each passing mile it was getting perceptively colder and I was finding it necessary to pull additional woolens from the sea bag — sweater, field jacket, lining for field jacket, outer trousers, lining for outer trousers.

I wanted to do a story about all this before my fingers were too cold to type. With my elbows pinched to my sides (it's crowded aboard a MATS Super Constellation with all those clothes and seventy-seven men — mostly scientists and Navy personnel — inside them), I started my lead: "This is being written at 18,000 feet . . ."

That's as far as I got. My Italian-made portable was longing for the sunny shores of the Mediterranean. It had, in short, become too numb to move — or at least space after each letter. . . .

Actually, typewriters are not the problem I may have made them seem. Only two days before my departure from the Naval Air Facility at McMurdo, which is the main American base in the Antarctic, some 2,300 miles south of New Zealand, I was ready to forgive my portable from freezing up on me.

I took it down from the rafter where I had disgustedly placed it in the press hut, and it worked beautifully. I nearly finished an entire story about its remarkable recovery before it grew numb again and quit spacing. The trouble was quickly diagnosed: I had moved too far from the hut's diesel fuel heater, which had kept the machine warm on the rafter. . . .

I wrote (on a borrowed typewriter) a spot news

story about a big plane crash on the ice runway at McMurdo, and passed it on to Navy communications for immediate transmission. It arrived at my paper six days later.

It wasn't the Navy's fault that it took so long to reach its destination. Navy communications in the Antarctic, as in all radio transmission, are at the mercy of the weather — particularly the weather on the sun. . . .

Sending stories back by air has its advantages, but soon after I arrived in the Antarctic a series of blizzards blew in that kept all incoming and outgoing planes grounded for a week. During this period I wrote a package of stories which I sent off on the first plane out after the storm. There were enough stories to tide me and my paper over before the next blizzard struck.

Although radio, planes, typewriters and cameras unfortunately don't have a magical ability to adjust to Antarctic weather, human beings do become acclimated.

I noticed this not long after my arrival in the Antarctic. I no longer felt the need for much of the thirty pounds of cold-weather gear, which the Navy issued our party of ten newsmen at Christchurch, New Zealand — advance headquarters for American activities in the Antarctic.

My ears got acclimated to the extent that I didn't have to go around with the flaps pulled down on my fur-lined cap. And though my bare fingers would still get cold taking pictures, I found little use for the issued pair of massive mittens that could be worn over two pairs of gloves.

Being a science writer, I was interested in exploring the reason behind this acclimation. . . . The Antarctic isolation provides scientists with a natural laboratory for studying men's reactions to cold and detachment — the sort of conditions men in space are likely to encounter.

CASE NUMBER FOUR

During the winter of 1964, David Perlman, science writer for the *San Francisco Chronicle,* left his desk in the Bay City news room and joined an expedition of 50 scientists going to the Galapagos Islands, those lands that Charles Darwin had visited a century earlier while aboard the *Beagle*. The Galapagos International Scientific Project, of which Perlman was a full-fledged member, was sponsored by the University of California and the National Science Foundation.

Since the reporter did assigned scientific chores as well as write news stories, his life was not always uncomplicated as he tells in the following narrative, which he wrote for the *NASW Newsletter* (September, 1964):

The idea was to write science, for a change, where it was really happening — not out of the journals, nor out of the annual meetings, but out in the field, where you never know from one minute to the next what's likely to happen. It was also a marvellous opportunity to develop a personal sense of intimacy with scientists from six nations and a dozen disciplines . . .

We travelled down to the Galapagos crammed into troopship quarters aboard an old freighter that sails the Pacific as a training ship for 300 Merchant Marine cadets from the California Maritime Academy.

In the black, bleak, cactus-and-jungle cluster of volcanic islands 600 miles off the coast of Ecuador we travelled on anything that moved: burros, feet and the single jeep ashore; ancient sailboats, native canoes, American helicopters and an Ecuadorian patrol boat by sea. . . .

Covering the scientists was a day-and-night affair, fanned out as they were over 23,000 square miles of ocean. In the daytime I worked with them; at night, while we squashed scorpions, shook centipedes from our sleeping bags on the rocky beaches, or scratched our fire ant bites together, we'd talk and talk. Dawn was a good time to write pieces — more than 30 of them — which I shipped to the *Chronicle* via Navy radio through the Pentagon, Navy seaplane through Panama, a drunken tuna boat skipper through Acapulco, or an American Embassy plane through Quito.

I even did a few scientific studies of my own . . .

The expedition research I covered was delightfully basic. Some was taxonomy pure and simple; 200 new lichens, 15 cacti, six Galapagos weevils. Some was be-haviorism; how do marine iguanas nest, or flamingos copulate? Some was atomic-age stuff; paleo-magnetic analysis of lava, and geiger-counter studies of iodine-131 uptake in lizard thyroids.

All of it was profoundly related to the Galapagos as a living laboratory of evolution, where Darwin pioneered and we followed. . . .

Reading the journals will never be the same for me, now that I'm back.

CASE NUMBER FIVE

Space exploration has been one of the spectacular and glamorous assignments for science writers since the late 1950's. Such news has been assured front page display and it is fea-tured with banner headlines and live broadcasting, always a satisfying experience for any newsman. Yet there is a continu-ing series of events that a smaller number of reporters has to cover, the run-of-the-mill (if that description applies to any of the astronauts' doings) activities, week in and week out, at the Manned Spacecraft Center at Houston, Texas, and around the country.

One of the reporters who has had this assignment is Harold R. Williams, aerospace writer for the Associated Press in Houston. He reminisced on working with the space astronauts in an article in *The Quill* (October, 1964):

To shield the astronaut from interruptions during the week, the Manned Spacecraft Center has set Friday only for interviews and photographs. This rule has caused the space center some mammoth headaches, but as far as I know it has not been broken. Foreign newsmen are frequent visitors and the astronauts are the big attraction for them.

Getting an astronaut interview isn't an easy assignment either. You make your request to the PAO [Public Affairs Office] early in the week, saying which one you want to talk to or take a picture of. The request is relayed to Al Chop, deputy affairs officer, who calls the astronaut officer for an appointment. Usually it isn't until Friday morning that you know if you can interview the busy space trainee. Friday also seems to be the day the astronauts attend long staff meetings ("always important"), leave Houston for a staff meeting someplace else, or catch up on their flying ("he has to keep his flying proficient — you know").

Once the interview is under way, the space man usually talks freely on any subject. He is articulate, intelligent and interesting. He discusses his family, his job, his feelings and beliefs, his hobbies, all with ease. But once away from the MSC office, he and his family are shielded by contracts for "exclusive rights to personal stories" by Time, Inc. and World Book Encyclopedia Science Service.

These contracts, which pay the astronaut and his family $16,250 a year — in most cases more than their annual salary, give Time and World Book the right to take photos of the family barbecuing or playing ball in their back yard, or just being together as a family. Other newsmen cannot get this type of photograph or story. The contract states that if another news medium wants an interview with the astronaut's family, or a photograph, permission must be given by World Book or Time.

Mrs. Betty Grissom, wife of Major Virgil (Gus) Grissom, who is spending every waking hour readying for his space flight in the first Gemini manned mission, was reported as saying, "I wish I had a contract so I could see my husband."

Because of the great news interest the American public

has in the astronauts, their training activities are closely covered. I have traveled with them to Panama for jungle survival training; to Arizona to look at the moon from Kitt Peak, and to tramp over jagged lava rocks near Flagstaff.

To keep newsmen at a minimum at some of these activities, a pool has been organized consisting of one black-and-white photographer, one color photographer, one movie photographer, and two writers, one from the Associated Press and the other from United Press International.

Everything written or photographed [by the pool representatives] is given to MSC for distribution to other media. This setup hasn't been without headaches, either. When astronauts came into an area, the local newspapers, radios and television want to do their own coverage, which is understandable. In some cases when the demands reach major proportions, the space center has bowed, and allowed everyone to participate.

CASE NUMBER SIX

Shortly after 6 P.M. on Tuesday, Oct. 5, 1965, the White House press secretary told reporters that President Lyndon B. Johnson was to enter Bethesda Naval Hospital for removal of his gall bladder the coming Friday morning. The news broke without warning, as far as Washington reporters were concerned, and how they reacted illustrates what can happen when stories with medical implications break through on to the front pages of the country's newspapers and as live broadcast "bulletins."

At the Associated Press' Washington bureau, Frank Carey, science and medical news specialist, checked in around 6 o'clock to leave a story to be printed the following day on the First International Conference on Desalination — itself a brainchild of President Johnson — which Carey had attended Tuesday. Then he took off with his wife for a Tuesday night cocktail party that some of the Conference members were having.

The bulletin on the forthcoming Presidential operation came into the AP office just about 10 minutes after Carey had checked out. Almost immediately after the news had been transmitted, the New York AP office was asking for an interpretive sidebar on what the medical aspects of the announcement were. Marvin Arrowsmith, AP news editor in Washington, replied that Carey had just left the office but he would try to locate him. Meanwhile, Arrowsmith suggested that New York try to flag Alton Blakeslee, AP science and medical writer who lives on Long Island.

Blakeslee had heard the news of the impending operation through a news broadcast and already was checking reference books at his home when his office telephoned him. As he explained it later, "Three books supplied the information needed, and this story was written within an hour." It moved promptly on the AP wires and was printed in many morning newspapers with the following lead:

> NEW YORK — President Johnson is now the victim of a trouble which may, at some time or other, affect one in ten Americans.
>
> His gall bladder is acting up, apparently from "stones" forming within this little-known organ of the human body.
>
> The gall bladder is a pear-shaped sac sitting under the liver just to the right of the stomach. Its job is to store and concentrate bile, an alkaline liquid manufactured in the liver. . . .

And the story continued along this general line of background information for enough to fill more than half a newspaper column in any AP newspaper that wanted to use it.

Carey, meantime, as he later told the story, was just working into his second drink at the Washington Hilton cocktail party when he happened to overhear the individual next to him say something like, "I just heard on the radio coming down that Johnson's gonna have his gall bladder out on Friday."

Carey asked his neighbor to repeat, then he checked his office by telephone and the switchboard operator said the news editor had been trying to reach him. After he arrived at his Washington office, Carey called Blakeslee and they decided that since New York already had Blakeslee's story, Carey would write his for the AP afternoon papers and radio/television stations to use on Wednesday.

What happened from then on was related by Carey as follows in the *NASW Clip Sheet* (January, 1966):

> Even from the meager available evidence in the White House announcement — especially the month's time-lag since his attack in Texas — it appeared to me that the President's gallstones must be confined to the gall bladder itself, or at worst, to the cystic duct. I checked this out with a surgeon friend and then wrote the piece . . . pitched from the angle that the LBJ brand of gall bladder disease was probably the less serious of the two varieties. And so, it held up — although, as you well know, ureter and kidney stones were extra added attractions.
>
> My surgeon friends continued to be a great help throughout the days that followed up to, including, and following the operation. I'd call them right from the hospital press room, getting a good deal of stuff to add to what [Presidential Press Secretary William] Moyers related to the reporters. My only regret was that, at their request, I could only refer to them as "Doctors not associated with the case said . . ."
>
> I'd say my original piece was ready a couple of hours after my hasty retreat from the cocktail party, but we held it up for moving in the PMS [afternoon] cycle.

The following spring Carey won the 1965 Journalism Award of the American Society of Abdominal Surgeons for his articles on President Johnson's gall bladder operation.

Nathan S. Haseltine, *The Washington Post's* medical reporter, was at the same cocktail party that Carey was attending and he returned by cab to his office. An experienced re-writeman already was working up background information to add to the story that Haseltine was going to type out. This background information told the President's medical history of a previous kidney stone removal in 1948, his 1955 heart attack, and even a 1964 toothache which necessitated filling an upper molar.

Haseltine called a urologist acquaintance to doublecheck his information but, as he explained, the conversation took 15 minutes of the half hour he had until the deadline for the *Post's* first edition. He wrote seven paragraphs when the copy desk called, "Time's up." These paragraphs, plus the additional background from the rewriteman and a diagram showing location of the gall bladder adapted from a medical reference book, all ran in the early edition. After going out for dinner, Haseltine came back and leisurely wrote an additional seven paragraphs for the later editions of *The Washington Post*.

John Troan, then Scripps-Howard staff writer in Washington specializing in science and medical news, recalled his experiences this way:

There was I, just beginning to enjoy the first evening of relaxation in more than a month [at a cocktail party], when some newsmonger sidled up to me and said: "You know what all the excitement has been about at the White House today?"

I thought he was being facetious so I replied irreverently: "Yeah, Lyndon's going to announce he's resigning the Presidency to become Pope."

"But I'm not kidding," says this guy. "The White House has just announced Lyndon's going into the hospital to get his gall bladder out."

So off went the spigots; I downed my Scotch-and-soda, checked the office by phone, scooped up some hors d'oeuvres, which I gulped without chewing to sandbag my ulcer, and raced out for a cab.

By the time I reached the office, the two men we had

at the White House (a Latin American expert who was there in event of a Cuban-refugee announcement and a Far East expert who was there in case of a Viet Nam development) had arrived with copious *medical* notes on LBJ.

I took the notes, pulled out my more copious file on gall bladders and gallstones, checked a couple of gall experts supplied by the D.C. Medical Society, and rapped out a story. We were off and running within an hour — so our morning papers could ride the story too. The story meant to be a feature-type sidebar; not the main story.

Next day I spent at the White House bird-dogging the President, attending a special briefing by his doctors plus a briefing by White House Press Secretary Bill Moyers, and obtaining additional information through personal inquiries to Moyers. Result: A story on how the President prepares for an operation by running himself (also, the press corps) ragged.

Meanwhile, I arranged for a phone to be installed for us at Bethesda Naval Hospital — for use during the President's stay there. The night LBJ entered the hospital I double-checked the phone connection, punched a Telex to make certain I could contact my office that way if the phone went on the blink, cased the parking lot, etc., and went to bed before Lyndon did.

I reported to the press room at the hospital before the President was to be operated on and remained through the day — until about 5 P.M. — to work up a story for the following afternoon. Again, the running story was outside my bailiwick.

As soon as it was disclosed the President had had a ureteral stone removed, as well as his gall bladder, I headed for the medical library at Bethesda Naval Hospital to brush up on urology. I skimmed through five textbooks, plus a number of pertinent journal articles to which I was steered by Index Medicus, and wound up with a list of 11 questions I wanted answered. I submitted the questions to Moyers — who read the answers to all hands at a late-afternoon press briefing. He also read answers to questions submitted by other newsmen, of course, so I could pick up some useful information that way too.

Then I went to the office to do a story on LBJ being a "stone former."

The story wasn't quite complete because nobody told us until the next day that Johnson also had a stone embedded in his kidney, which remains there. I honestly think Moyers was as surprised as the newsmen when the doctor mentioned this stone. However, I am convinced the White House (including Moyers) knew the President definitely had a stone in the ureter before the operation was booked but withheld the information. Why, I am not sure. They told us it was because the doctors weren't positive enough to talk about it beforehand. But I'm convinced it was because the White House feared an unfavorable public reaction if it was disclosed simultaneously that LBJ had both a gallstone and a ureteral stone. Fears of public panic may have warranted withholding of this information; at the same time, it made me question, in my own mind, how much faith I could put into subsequent White House statements about the President's health. There remained, thereafter, the nagging doubt, the feeling that maybe something else was being withheld. Sudden revelation, the next day, of the stone embedded in the kidney only served to reinforce this feeling. I am not criticizing the White House staff for this — because I don't know how I'd act were I in their shoes. I am simply relating my experience and my reaction.

I kept tabs on the President's progress after this but, in the absence of any truly significant development, I wrote no more stories because the regular wire services were covering the case adequately enough for our papers.

A number of science writers on newspaper staffs across the nation called up local physicians to obtain "local angle" items about the President's operation. However, others felt that as long as the wire services, especially Associated Press and United Press International, carried adequate stories there was little reason for them to get into the act also. One reporter, Mildred Spencer, medical writer with the *Buffalo Evening News*, 1965–66 NASW president, and 1965 winner of an American Medical Association newspaper writing award, explained why she did not write any special coverage for her paper:

> I have some real reservations about a certain type of "local" story that is done in cases like this. The reporter goes to an eminent local physician and asks him how such surgery is handled and gets *his* answer. This may not necessarily be the way the President's surgery is going to be handled, and all of the prognosticating looks pretty silly afterwards. For example, local physicians in Buffalo, and many other places, do not restrict the diet in any way after gall bladder surgery. Yet wire stories said that the President's diet *was* being restricted. Presumably there are reasons for this which do not always apply — but no local physician of any repute would attempt to give a "prescription" for treating a patient he had not seen.
>
> I would hope NASW members were not guilty of some of these attempts to master-mind a story at long distance, but I have seen a number of this kind of story in the past.
>
> If you're not going to do it this way, then you can always go to the doctors who are going to do the work, but I think it an imposition, myself, for every reporter in the country to be calling a doctor and asking him the same questions, just to write a story saying "The President's doctor told ME — "
>
> It seems to me that the function of a local reporter in a case like this is to read the official reports sent via wire or Washington Bureau with a sharp eye, make the proper queries to sharpen a point or correct an error, etc. It is also quite legitimate, of course, to do a story on the gall

bladder, what it is, how it works, how we can get along
without it. But I don't really see any real reason why *I*
should do this when it is well done by a wire service.

CASE NUMBER SEVEN

That broadcasting newsmen assigned to science coverage have their own problems — and sometimes their unexpected rewards and suspenseful moments — has been described by Jules Bergman, science editor with the American
Broadcasting Company and the first television staff member to
gain that title. Writing in *Television Quarterly* (Spring, 1963),
Bergman told of his experiences in December, 1961, at the
United States' first underground nuclear blast to test the peace-
ful uses of atomic energy:

It was before dawn, about 5:30 A.M., on the New
Mexico desert, some 35 miles from Carlsbad. The event
was called Project Gnome, and it was the first shot in
Project Plowshare. A raw wind was pumping 10° air
through the thin raincoat I was wearing. (Someone had
told me it was warm and sunny in New Mexico!) I'd just
finished a film interview with one of America's most famous physicists who assured me that only a slight shock
wave would be felt and that no radiation could possibly
reach the surface.

Ground Zero was some two miles away, and as the
countdown reached zero nothing at all seemed to happen. The famed physicist was in a helicopter orbiting the
blast area. The chopper turned tail at about the moment
the shock wave lifted me, the camera, and the ground
about 12 inches into the air. Through my field glasses,
I watched the desert floor at Ground Zero heave some
four feet into the air, then fall back abruptly. Just as I
was about to tell the cameraman to stop the camera,
another physicist (somewhat less famous) tapped me on
the shoulder.

"Leave the camera running," he muttered quietly.

"You'll get some interesting film." As I turned back to the blast site, a thin plume of steam thickened into a geyser-like flow out of the ground. "It's venting!" somebody shouted. As usual, no official had any explanation, but many phones were picked up by many officials, and suddenly there were warnings that we would have to move out of there like lightning or not be allowed to leave at all.

The radioactive cloud (of a low order, it later turned out) moved in our general direction, then veered off to the side. Just as I was about to pick up my film, state troopers closed-in from all directions. "No one leaves," the order went out. And there wasn't even a working phone to call in the story for the ABC radio network. I made a few experimental sallies at the roadblocks, but the troopers heavily outnumbered me. I did glean, however, that I could leave by the road to the north — it was only 100 miles back to Carlsbad that way. I also found out that if I were daring, there was a dirt road across the desert.

So off I went, with another reporter, one eye on the dirt road, the other on the radioactive cloud as it pushed steadily toward us. After two hours, we had gotten completely lost; a few Mexican farmers offered no help; and we were running out of gasoline. Desperately, I took the less likely of two dry-gulch trails and ended up at an old potash mine shaft. It even had a phone, though no one in New York believed me when I called in. After borrowing some gas, I headed off toward Carlsbad, clutching the can of "hot" film (hot indeed, possibly). We passed two more roadblocks, and the troopers waved us on when we told them we were merely lost tourists. "Who me? A reporter? Not a chance. . . ."

Reaching Carlsbad, I found I had a "hot" car; we both got a quick radioactivity count and passed. I had driven through the radioactive cloud, but it had dissipated by then. After battling a snowstorm through the mountains to get out (I had hit a dust storm flying myself in), I reached a jet at El Paso at 4 A.M. and got the film to New York. Meanwhile, back at the press site, everyone else was still trapped.

The moral of this story, other than to always carry a Geiger counter, warm clothes, spare gas, and a good compass, is somewhat vague. But it is roughly this: Scientists can be wrong, too, and part of the adventure of my job is being there when experimental tests do go awry. And even with the degree of failure, the shot — to determine if nuclear blasts deep in salt caverns could be harnessed to produce heat and thus useful energy — was largely successful, though there are a lot of newsmen who remain unconvinced.

9

SCIENCE NEWS "CONSUMERS"

Just before 6 P.M. on Thursday evening, March 17, 1966, the National Aeronautics and Space Administration's Manned Spacecraft Center at Houston received a message from the then-orbiting Gemini-8 capsule; the message foreshadowed crucial difficulties. It said:

> Well, we consider this problem serious. We are toppling end over end but we are disengaged from the Agena. . . . It is a roll or nothing, we cannot turn anything off. Continuing in a left roll. . . . Stand by.

As one NASA official confessed later, "It looked like the whole ball of wax was gone."

After a short delay, NASA gave out a public announcement and the three national television networks announced that the Gemini flight was in trouble. The Columbia Broadcasting System decided to continue news broadcasting after the regular 7:30 P.M. EST closing time, pre-empting, ironically, a scheduled entertainment program called "Lost in Space." The American Broadcasting Company and the National Broadcasting Company started their scheduled entertainment programs, "Batman" and "The Virginian," respectively.

By the time the early splashdown decision was made official at 7:44 P.M., ABC and NBC had joined CBS in covering the Houston anxiety.

ABC officials in New York reported that they promptly received several hundred complaining telephone calls during the approximately 12 minutes that had been pre-empted from the entertainment program for live news coverage of the Gemini flight's distress. According to them, one "Batman" addict telephoned from Cleveland to complain, "Me and the kids can't follow the plot with all these interruptions. Get with it, will ya?" Another "Batman" devotee was quoted as saying, "I was plenty mad because the Gemini stuff had nothing to do with 'Batman.' I missed out on Batman finding the clues and I missed out on the fighting. They should have flashed their news across the bottom in print instead of interrupting the whole show."

In Los Angeles, where a scheduled showing of "Batman" was knocked off by full-time Gemini coverage at the later hour, more than 3,000 telephone complaints were received and the station rescheduled the entertainment program for belated showing another night.

In New York City, NBC received a gripe from a Chicago listener who said, "I'm spending my money on this call because it's disgusting what you people are doing. We tune in to see 'The Virginian' and we get this dopey Gemini stuff."

And CBS, which had continued broadcasting the Gemini news immediately after its regular program was due to end, received complaints from viewers who protested scrapping the fictional show "Lost in Space" for the real-life plight of the Gemini astronauts.

One New York City network executive said, "They accused us of scrapping their favorite programs and putting on trivia."

These incidents demonstrate that audience interest in science news can be remarkably low, even when it may be, as one Gemini-8 flight newspaperman wrote the following morning, "probably the most tense moment in the history of the United States manned space flight." Such suspenseful news, said to have cost the networks an estimated $3,000,000 in cancelled schedules, should have met many of the requirements of "hot copy" — but for some people, apparently, it failed in competition with television entertainment.

Several studies of attitudes toward science news have shown that there are always some individuals who are not attracted, just as there are undoubtedly some who have no flair for reading, viewing, or listening to sports, society news, politics, or international affairs. But such surveys as we have also indicate that a majority are attracted to science, technology, and medicine, and that a considerable proportion of them would like to have more of this news included in newspapers, magazines, radio, and television.

The response to science news and information, as pointed out in Chapter 4, runs the gamut from complete ignorance on the part of those who do not read, view, or listen to the items already there to high interest on the part of those who absorb the technical details and move into action. Thus the range is from the science-blind to the science-activated. What we are concerned with in this chapter is those who absorb and, to some degree, those who act.

Just who are the science news "consumers"?

A 1957 survey of 1,919 adult Americans by the Survey Research Center of the University of Michigan for the National Association of Science Writers and New York University under a grant from the Rockefeller Foundation provided some of the characteristics of those who reported they "consumed" the science and medical news of the mass media. Some of the more noteworthy conclusions of this study included:

• Three-quarters of the total sample (76 per cent) could remember a recent, specific, science or medical news item and thus qualified as true "consumers."

• Men tended toward science items, women toward medical news.

• Those who recalled science and medical items in the mass media tended to be better educated, had higher incomes, and were consistent media-users for all sorts of news.

Since staff members at the Survey Research Center wanted to obtain a "conservative" measure of how many people "consumed" science news, they constructed questions which made the greatest possible demand on respondents: they asked them to recall actual science items from the mass media. Almost a quarter (24 per cent) could recall no specific story about science or medicine. At the other end of the spectrum, one-tenth of one per cent — two individuals in the 1,919 sample — could recall both science and medical items from all four media: newspapers, magazines, radio, and television.

Looking at the science audience and at the medical audience individually, we found that the latter was larger with 69 per cent remembering some medical story, 52 per cent of the total sample recalled science news. Approximately one-quarter (26 per cent) of the sample respondents remembered something about non-medical science news from a single source, while more than a third (36 per cent) recalled a medical item from one source.

A breakdown of multiple media sources for science and medical news recalled showed:

RECALLED ITEMS FROM:	MEDICINE	SCIENCE
FOUR MEDIA	1%	1%
THREE MEDIA	8	8
TWO MEDIA	24	17
ONE MEDIUM	36	26
NO MEDIUM	31	48
	100%	100%

(Sample size = 1,919)

Looking at the media individually, we find newspapers the favorite channel among all respondents for getting science and medical news; 64 per cent remembered at least one specific item they had seen in print. Television was next with 41 per cent; magazines, 34 per cent; and radio, 13 per cent. In view of increased television coverage of science and medicine — this survey was made shortly before the 1957 Sputnik — I would speculate that television has gained interested audiences more rapidly than the other media, although it is unlikely that television viewing has replaced newspaper reading as the main mass-medium source of science and medical information. Unfortunately, there have been no extensive national surveys to test this assumption.

On the basis of each medium's individual audience, the figures showed that seven out of ten (71 per cent) in the newspaper audience recalled at least one science or medical news item. Approximately half of both the magazine (52 per cent) and television (47 per cent) audience and three out of 20 (16 per cent) of the radio audience recalled one item. The television audience, however, was large and the magazine audience, smaller.

Asked about content, those who recalled science stories tended to emphasize the technological aspects, such as military uses of atomic energy and other defense adaptations of scientific and engineering developments; consumer, industrial, and agricultural applications also had prominent mentions. When medical items were recalled, the emphasis was on such few well-publicized diseases as the major causes of death — heart trouble and cancer — and on infantile paralysis, which was much in the news during the late 1950's because of the massive testing of vaccine across the nation.

The survey showed that men and women varied in their reading and use of science and medical news. This is shown in the following tabulation from the NASW-NYU findings:

	MEN	WOMEN	COMBINED
RECALLED MEDICINE	66%	71%	69%
RECALLED SCIENCE	61	45	52

This same predilection of men for non-medical science and of women for medicine has been displayed in the Roper and Gallup public opinion polls over the years. For instance, men scored higher on tests asking them to identify Albert Einstein, explain radioactive fall-out, and so forth, while women scored higher in explaining tranquilizers, the relationship between lung cancer and cigarette smoking, and the symptoms of Asian flu.

To interpret these trends, I would guess that wives and mothers, more concerned with the health of their families than the men in the households, sought out information that might help them better fulfill these responsibilities. On the other hand, husbands are usually the family "gadgeteers" with an interest in do-it-yourself mechanical hobbies and other such activities. Both men and women were chiefly attracted to science/medicine news because they wanted to keep up with things that were going on. However, slightly more men found science and medicine "interesting" and slightly more women found such information "helpful to me in everyday life." Thus, the males tended to stress intellectual curiosity while women favored the utility of learning about science and medicine.

Education appeared as an impressively influential factor favorable to "consumption" of science and medical news. A breakdown of the percentages of "recalled stories" in these areas by education levels in the NASW-NYU study showed:

	MEDICINE	SCIENCE
GRADE SCHOOL	46%	29%
SOME HIGH SCHOOL	72	49
COMPLETED HIGH SCHOOL	83	67
COLLEGE	90	81

Impressive as these statistics are, a closer look at those who took physical and biological science courses in school or in college showed that it was not higher education but course work in the sciences which apparently kindled a life-long interest in things scientific and played the more decisive role. This was graphically illustrated in the following percentages:

EDUCATIONAL BACKGROUND	PERCENTAGE WHO RECALLED SCIENCE OR MEDICINE FROM AT LEAST ONE MEDIUM	PERCENTAGE OF TOTAL SAMPLE IN GROUP CITED
GRADE SCHOOL	54%	36%
SOME HIGH SCHOOL		
NO SCIENCE	75	12
SCIENCE	85	9
COMPLETED HIGH SCHOOL		
NO SCIENCE	80	4
SCIENCE	91	21
COLLEGE		
NO SCIENCE	86	1
HIGH SCHOOL SCIENCE ONLY	92	6
BOTH HIGH SCHOOL AND COLLEGE SCIENCE	99	11
		100%

(Sample Size = 1,919)

Notice that high school graduates who took science courses could recall items more readily than those who attended college but took no science courses in either high school or college — and nearly as easily as those who took no science beyond high school. With practically all high school students now required to take at least some science courses, and with most reputable colleges requiring additional work in this field, the prospects for increased science news "consumption" in the near future indeed appear bullish.

Elmo Roper and his associates polled a cross-section of adult Americans to ascertain their intellectual and cultural involvement and reported the results in *Saturday Review* (May 14, 1966). When respondents were offered a list of subjects and asked to name those in which they had "a great deal of interest," the subject of "Science" got 20 per cent, well behind religion (49 per cent); sports (47 per cent); music (46 per cent); politics and government (40 per cent); but ahead of literature (19 per cent) and art (13 per cent). When those who had been to college were tabulated separately, the percentage for "Science" rose to 36 per cent. Again demonstrating that a college education often — but not always — stimulated interest in science and science news "consumption."

In their final report on a mental health project at the University of Illinois in 1960, Drs. Jum Nunnally and C. E. Osgood said that the chief islands of ignorance about mental health and illness were among the older and the least educated people. When education was held constant, little effect attributable to age remained, thus indicating, as just cited from the NASW-NYU study, that modern education's emphasis on science and public health was undoubtedly a more influential factor than age alone.

Why is there this apparent linkage of readership and knowledge with education in science?

Speculation might center on three factors:

1. Those who study science have some of the background needed to understand it when they use the mass media. In other words, science courses probably build a "science literacy."

2. Science education provides a bridge for relating scientific developments to public affairs and is a step towards better social conditions.

3. Science courses probably stimulate a life-long interest in and a respect for the subject.

Regarding this last point, self-screening also may play some role, especially if science courses are electives which students may choose or reject. With the "new science" and "new math" curricula required for college preparatory high school students, with some science work required by most colleges and universities as prerequisites for admission, and with the mounting millions of teenagers seeking to enter college every year, this self-screening factor of the past few decades may dwindle into insignificance for youngsters now being educated.

Income, a factor not often unrelated to educational background, was an index for measuring the use of science and medical news from the mass media. The NASW-NYU study showed the following percentages for recall of science and medical stories by income gradations:

	MEDICINE	SCIENCE
LESS THAN $1,000	34%	21%
$1,000–1,999	44	29
$2,000–2,999	60	40
$3,000–3,999	67	49
$4,000–4,999	69	57
$5,000–5,999	78	59
$6,000–7,499	83	65
$7,500 AND OVER	87	71

Avid readers of science news seemed to be eager "consumers" of most other areas of news, especially of medical, local, national, and foreign affairs news; they were somewhat less interested in crime, sports, and society. Avid readers of medical news tended to be less enthusiastic about such other news but they displayed some inclination toward science, local items, and people-in-the-news or human interest features. Thus, science readers were more attracted toward the intellectual slice of the news while medical news "consumers" leaned toward a more personalized and local perspective. And, according to those who supervised the University of Michigan field interviews, "Simply to interpret the two groups by contrasting masculine and feminine interests is not an adequate explanation."

Although the science and medical news "consumer" may be in any age bracket, he is more likely to be among the young and the middle-aged. It is probable that this simply reflects the increase in the opportunities to study science in high schools and colleges in more recent years since, as pointed out previously, one becomes sensitized to science in the classroom.

Summary: In the light of existing studies, then, the typical science and medical news "consumer" is likely to be younger rather than older; his family income is probably well above average; he probably studied science in high school and, if he went to college, took additional courses there. Among those attracted to medical news, women are more likely to predominate than men, and, among those attracted to non-medical science, there are probably more men than women.

The over-all philosophy of the adult who is a whole-hearted consumer of science and medical news was pictured this way by Dr. Robert C. Davis in a report he wrote on the 1957 field study, *The Public Impact of Science in the Mass Media* (Survey Research Center, University of Michigan, 1958):

> He is more attuned to the large world around him; his vista is more cosmopolitan than local. His interests range from the immediate community to the world scene. His concern with the broad picture is reflected in his reasons for reading science: he wants to keep up with the world and he wants to know how science will shape his destiny — and his chances of survival. . . .
>
> He expresses his feeling about science news in terms of a desire to see science-in-context. It helps him make sense of his world as well as to function in his personal life. He sees science as beneficial, and assesses its impact on society in terms of improving our way of life. Although he may be concerned with possible bad consequences of scientific discoveries, such as atomic warfare, he does not blame scientists for these consequences.
>
> Rather, scientists are viewed as diligent, educated, intelligent people whose hard work is motivated not by self-interest in the economic sense, but by the intrinsic interest in the endeavor called science. He sees the scientist as different from the average person, but dedicated to constructive ends.
>
> Unlike a minority of his fellow citizens, the science con-

sumer does not feel that the deviance of scientists is a worrisome thing. He is less likely to be concerned with the possibility that science is shaking the traditional and moral foundations of society. And he does not feel that scientists are odd, prying or irreligious. . . .

He tends to view the world from a perspective similar to that of science. He is optimistic as to the range of problems science can tackle; he feels the world to be not mysterious chaos, but to be knowable and orderly.

In terms of his view of his social world he is also more inclined to see it as manageable and essentially benign.

All in all, the science consumer confronts his world with a general desire to know and understand it. The world is, in a broad sense, not overwhelming or threatening, but an area in which to act and master, either by his own endeavors or by vicariously participating in the enterprise called science.

Several studies of high school and college students during the past decade* showed that many of their attitudes toward science and scientists reflected a blend of those held by their parents and their teachers. Thus they were ambivalent about the scientist as a "good" or "bad" person although a considerable majority of them favored positive images rather than negative.

A study of between 10,000 and 18,000 teenagers by Dr. H. H. Remmers and D. H. Radler, both with Purdue University when they published *The American Teenager* in 1957, confirmed this point. When asked their opinions of the statement, "There is

* See especially the following:

Margaret Mead and Rhoda Metraux, "Image of the Scientist among High School Students," *Science*, CXXVI, No. 3270 (Aug. 30, 1957), 384–90.

H. H. Remmers and D. H. Radler, *The American Teenager* (Indianapolis: Bobbs-Merrill Company, 1957).

David C. Beardslee and Donald D. O'Dowd, "The College-student Image of the Scientist," *Science*, CXXXIII, No. 3457 (March 31, 1961), 997–1001.

something evil about scientists," the teenagers replied as follows:

AGREE	7%	} 14%
UNDECIDED, PROBABLY AGREE	7	
INDIFFERENT	10	
UNDECIDED, PROBABLY DISAGREE	14	} 70
DISAGREE	56	

(Table does not total 100 per cent because some students did not respond to this question.)

High school seniors expressed outright disagreement more vigorously than their freshmen schoolmates, who had not yet taken science courses, by a ratio of 60 to 52 per cent. Those from families with higher incomes disagreed more than those from low-income homes by a ratio of 62 to 45 per cent. Those whose mothers had gone to college dissented most of all (74 per cent), compared with 50 per cent for those whose mothers had only a grade school education.

The same teenagers tended to disagree (53 per cent) with the statement, "Scientists, as a group, are more than a little bit 'odd,'" while 25 per cent agreed. The rest were indifferent or undecided.

Many scientists are well aware of the ambivalent, and sometimes anti-science, attitudes held not only by teenagers but by adults as well. They certainly should be concerned, not only to avoid being overwhelmed by a negative public reaction, but also to prevent, hopefully, a worsening of the social climate.

Dr. René Dubos, professor at Rockefeller University and author of science books sufficiently popular to appear in paperback format, commented as follows in the Winter, 1965, issue of *Daedalus:*

The immense majority of the lay public shows by its reading habits that it sharply differentiates between science and non-science; this differentiation also appears in the fact that concert halls and art museums have more popular appeal than science exhibits. The "two cultures" may be an illusion, but in practice science is still regarded in our communities as a kind of foreign god, powerful and useful, yes, but so mysterious that it is feared rather than known and loved.

It is healthy to acknowledge that scientists themselves generally behave like the lay public when they function outside their areas of professional specialization. The student of plasma physics or of plasma proteins is not likely to select books on marsupials for his bedside reading, nor is the organic chemist inclined to become familiar with problems of population genetics. Most scientists, it is true, are interested at present in radiation fallout and in the hidden surface of the moon, but so are many members of the Rotary Club. . . . In brief, while scientists are deeply committed to their own specialized fields, they generally turn to non-scientific topics when they move outside their professional spheres.

In an effort to determine the intensity of the reading habits of those who said they regularly read science and medical news in their daily and weekly newspapers, the Survey Research Center investigators inquired how much the individual respondents read. Results for newspaper readers showed:

	MEDICAL NEWS	SCIENCE NEWS
READ ALL	41%	30%
READ SOME	35	32
GLANCED AT	13	18
SKIPPED OVER	10	18
NOT ASCERTAINED	1	2
	100%	100%

(Newspaper audience = 1,751)

Reflecting the differing interests of men and women readers, the study showed that 51 per cent of the women and 27 per cent of the men reported that they read all of the medical and health news, while 39 per cent of the men and 23 per cent of the women said they read all of the science news. At the other end of the scale, twice as many men as women (14 to 7 per cent) skipped over news of medicine and health and almost twice as many women as men (22 to 12 per cent) by-passed science items.

Those who "consumed" all the science and medical news they found were more numerous than comparably eager readers of national politics, foreign events, society, and sports items. Local news and human interest stories both had larger audiences than science and medical news.

Not only did the individuals in the newspaper audience report extensive readership but they also said they wanted more stories in these two categories and were, in the main, willing to reduce coverage of other areas to obtain space for such expansion. The figures on those who claimed desire for more science and medical news were as follows:

	MEDICAL NEWS	SCIENCE NEWS
WANTS MORE	46%	30%
WANTS SAME	47	59
WANTS LESS	4	7
NOT ASCERTAINED	3	4
	100%	100%

(Newspaper audience = 1,751)

As might be expected on the basis of previous findings, those respondents with better educations and higher incomes were the most interested in expanding science and medical coverage in their papers. Predictably, women plugged most heavily for medicine and men for science.

Since newspaper space is not unlimited, the questioners asked what might be cut to provide increased science and medical news. Here are the results:

CUT NOTHING	32%	*Heaviest among those with grade school education.*
SOCIETY	21	*Predominantly male readers.*
CRIME	19	*Chiefly high school graduates and college attendees.*
SPORTS	17	*Heavily female readers.*
ADVERTISEMENTS	13	*Generally distributed, but slightly more males.*
COMICS	10	*Slightly heavier among those who went to college.*
POLITICS	6	
SCANDAL	5	
MISCELLANEOUS FEATURES	3	
OTHER, SCATTERED	9	
NOT ASCERTAINED	2	

(Totals to more than 100 per cent because more than one response was given.)

(Sample size = 1,919)

Not surprisingly, men were willing to cut society news to obtain the necessary space while women suggested reducing sports and crime news. Each sex was perfectly agreeable to curtailing those departments in which there was the lesser interest, as the following figures also show: 32 per cent of the males and 13 per cent of the women wanted to curtail society news, 20 per cent of the women and 12 per cent of the men said to limit sport space, 22 per cent of the women and 15 per cent of the men said to cut crime news, and 15 per cent of the men and 11 per cent of the women wanted to limit advertising.

Among those who said they read all the science stories that got into print, 31 per cent were willing to cut society news, 21 per cent each for sports and for crime, and 17 per cent for advertisements. Of the "Wants more" science news group, men again would cut down on society items (42 per cent) and women, to a lesser degree, on sports (24 per cent).

Respondents were also asked to rate the four mass media for completeness, interest, accuracy, and ease of understanding. Results gave high ratings for both television, the medium that combines both audial and visual presentations, and magazines. Both were ranked the same in all four categories except completeness, where TV was a close second to magazines. It should be kept in mind, however, that an audience that could remember at least one specific science or medical item from the medium involved were doing the rating. And, since newspaper readers were the largest group involved, the number satisfied with press reporting obviously included more individuals than the number who expressed satisfaction with the jobs being done by either television or magazines.

While general satisfaction ran high among the science and medical news "consumers," there was just enough uneasiness expressed to suggest that the powers-that-be might do a little soul-searching and take some second looks at the mass media's popularizations of these news areas. Avid users of such information from the news media are generally favorably and positively inclined, but what of the others?

To what extent do deadlines, stylized format, and limitations on time and space interfere with a more meaningful presentation? Can these barriers be eliminated without an upset of the operating procedures that have long and stubborn traditions?

To what degree are difficulties due to human elements—insufficient background on the parts of both newsmen and audiences, shoddy or hasty work, the complexities of translating the nearly untranslatable languages of science, technology and medicine?

10

VIEWS FROM THE LABS

By the mid-1960's, attitudes of scientists, engineers, and physicians toward popular reporting of their activities generally could be summarized as a "Yes, but . . ." reaction: an almost unqualified approval for most of the work of the professional science journalists, an occasional kind word for anyone assigned to interpret research and operations, and often a wistful hope that things would get better in the future.

Obviously, since there are all kinds of science reportage, various groups have vastly different reactions. However, those who meet the professional reporter most often have the kindest remarks to make. Individual scientists who have presented few papers or have little other contact with media representatives remember only the horrifying experiences of their predecessors who lived during the heyday of "yellow journalism." For instance, a study made during several conventions a few years ago showed that the overwhelming majority of the published scientists who were questioned had little basis for complaint about their own experiences; nevertheless they expressed a cautious restraint about the extent to which they could — or should — trust journalists. This reaction apparently had been handed down as professional folklore from one career generation to another, regardless of what actually had happened to the individual when he encountered newsmen. Some trouble also arose because of differences on the goals of popularization.

Predominantly responsible for a recent change in the attitudes of those who get extensive publicity has been an increase in funds allocated to research. And, to a considerable degree, the change is also due to a shift from financial patronage by the rich to support from the public, either in the form of government grants and contracts or in the form of money collected by such non-governmental but public agencies as the American Cancer Society, American Heart Association, and the National Foundation. And this has taken place regardless of whether the former patrons were individuals, corporations, or foundations.

Writing in *Modern Medicine* (March 30, 1964), Dr. Irvine H. Page, director of the Research Division of The Cleveland Clinic, pointed out:

> Make no mistake about it, today's science writer is a truly essential part of science and medicine. On him is a heavy responsibility. We may be able to decide what we would like, or ought, to do but whether it will be done is often in the hands of the members of the National Association of Science Writers and their ilk. Believe me, they are important friends to have and, even better, they are friends you will like to have. The "Fourth Estate" must now be included in the estate planning for the spiritual

and intellectual welfare of the body of science and medicine. Should I mention material welfare as well?

Commenting on the recent changes in scientists' opinions
of the mass media in June, 1961, Dr. Robert Brode, Department of Physics, University of California, Berkeley, California,
reported the following picture of contemporary attitudes to a
group of scientists, science writers, and journalism teachers in
Washington, D.C.:

> Today there is a changing attitude of the scientist to-
> ward the press. Some years ago if a scientist permitted
> close contact with the press and permitted them to pub-
> lish stories regarding his work, he was almost thrown
> out by other scientists. Now most scientists realize the
> need for the press. We seem to go from one extreme to
> the other in this relation. There is a place for scientists
> who are so concerned in their work that they do not spend
> a large proportion of time in public relations. They
> should be able to leave that to those who are interested
> in this field. There are scientists who are interested in
> public relations and who work well with the public and
> with the press. These scientists should be encouraged.

Critics of media performance in covering science, technology, and medicine tend to concentrate their attacks on three
different fronts:
1. Poor selection of material chosen for publication.
2. Journalistic operating procedures that tend to maximize
inaccuracies and distortions.
3. Inadequate or faulty training of mass media reporters.
A charge of "sensationalism" in handling news is probably
the one criticism most frequently cited by scientists. And they
can document this with enough incidents to make all but the
most calloused journalists cringe in embarrassment. Yet some
of the scientific and medical advances of the past several decades have, in fact, veered close to earlier science fiction. When
fiction of the past becomes reality, it tends to be spectacular —

no matter how the news is handled by the science journalists.

Physicians rightly complain that too many newspaper reports, especially of new drugs, have ballyhooed medical findings as "cures" for some fairly common disease. Cancer research frequently has been a ground for such reporting; actually hundreds of chemo-therapists can tell of a deluge of plaintive letters requesting the "miracle cure" after the mass media have told about a medical convention paper or journal article. Yet the public has as much "right to know" about such research, especially if its money is being used for financing, as it does about other governmental activities undertaken with public funds.

When I asked his opinion several years ago, Dr. Vincent du Vigneaud, 1955 Nobel Laureate for Chemistry and a professor at Cornell University Medical College, complimented medical reporting in the New York City papers but added this about general coverage:

> There is of course a tendency toward the sensational, and year after year the same disease seems to be cured. There is no question that the science writers are attempting to raise the level of their writings. I think it will be rather slow progress. The newspapers must recognize their responsibility. It is my impression that many times the science writer is blamed for things that really lie at the door of the scientist.

This same feeling by the non-medical scientist was demonstrated by an editorial in *Science*, a weekly publication of the American Association for the Advancement of Science and recognized as widely-quoted spokesman for that large group of individuals interested in science, its applications, and developments. Dr. Philip H. Abelson, the editor, wrote in the January 18, 1963, issue:

> It is true that the volume of news of science in daily newspapers is increasing. In Washington and New York, coverage is excellent; the writers are exceptionally com-

petent, and sometimes adequate space is devoted to their stories. In other parts of the country science reporting ranges from fair to downright mediocre, or there is none at all. Some good, authoritative material is provided by the wire services, but local editors butcher it with a heavy hand. . . . Science writers for the wire services, wanting their copy to be used, tend to seek the more glamorous items. With distressing frequency science-operators are able to flim-flam the science writers with news stories which excite the imagination but have no solid technical basis. Local editors are especially susceptible to these worthless baubles, which they run in preference to less exciting items of solid merit.

That the scientists and physicians and their lay sponsors share some of the responsibilities for faulty news reporting in the mass media was pointed out by a Nobel Laureate who asked that his name not be used. He wrote to me:

Present-day faults lie more with the medical profession and their lay sponsors than with the press. The major philanthropic institutions and foundations, as well as various branches within our own government, well realize the importance of publicizing medical research in terms of aiding and abetting fund-raising activities or requests for appropriations.

Unfortunately, the pressures deriving from such sponsoring agencies for news releases have led to the phenomenon wherein not infrequently scientific advances are announced first in the lay press. Thus, almost weekly we read of some important new "cure" for this or that disease. The press is not to be blamed for this situation but rather the scientist who succumbs to the pressures for the fund-raising components of his respective institution or foundation. The solution to this problem is not in sight.

The world-famous scientist just quoted and du Vigneaud, quoted earlier, both indicated the influence of public relations and the press release on science news coverage. This influence can be strong and, at times, must be if a small number of reporters is going to reach a large number of news outlets. However, this will be discussed in greater detail in Chapter 12.

Sensationalism need not be only inaccuracies, untruths, or excessive public relations. It may arise from misdirected emphasis and shifting facts out of context. Thus, a news item, magazine article, radio or television script may report factually what was said but this may be torn out of its perspective and its qualifications deleted, leaving readers, listeners, and viewers to draw an incorrect over-all impression.

Dr. Howard Burchell of the Mayo Clinic observed that, while one now rarely hears serious objections to what has been reported, medical scientists, when they do complain, are concerned with "the tendency for a newspaper article to magnify the importance of small contributions and seemingly to ignore the fact that progress usually is made gradually with many persons contributing to it."

To give another example, Dr. Edward L. Tatum, 1958 Nobel prize winner in medicine and physiology and a member of the Rockefeller University staff, has cited the "frequent tendency to play up and exaggerate the significance of a contribution out of all proportion." He referred to (1) the overuse of such words as "breakthrough," "major advance," and "key to life" when reporters are explaining matters of strictly secondary import and (2) those stories written as if the most recent findings were completely new, instead of simply based on, and continuing from, earlier work.

Science writers, especially those who have attained professional levels, are aware of the journalistic sins cited by Tatum. To illustrate, Victor Cohn of the *Minneapolis Morning Tribune* concluded a talk at the 1964 AAAS meeting with this tongue–in–cheek remark: "And now I must close. I may be missing three or four major breakthroughs."

However, it occasionally may be easier for the lazy or less competent newsman to pump up popular attention to a new development with adjectives rather than with a basic, general analogy or a translation of a scientific technical concept into language the man in the street can understand. Overplaying a sensational twist at the expense of happenings and topics — non-sensational but still extremely significant — creates an unbalance and thus propagates considerable misinformation, as more than one scientist and physician has explained. A little more effort by the reporter could have made it all much clearer for the readers or listeners.

Addressing the 1961 Pulitzer Prize jurors, Dr. Polykarp Kusch, Columbia University physics professor and a 1955 Nobel Laureate, spoke of the following problems in science popularization:

> I think that science is frequently badly presented to the public. . . . The public is bombarded by news of ever–new triumphs of science and fails to understand that even science has its limitations. Science cannot do a very large number of things and to assume that science may find a technical solution to all problems is the road to disaster. For example, not all the power of science can in the long run feed an ever–expanding population and the assumption that it will do so is dangerous.

> Science, in itself, is not the source of the ethical standards, the moral insight, the wisdom that is needed to make value-judgments; though it is an important ingredient in the making of value-judgments. Social, political and military decisions are made on grounds other than those in which science is authoritative. . . . I think a more cautious use of the phrase "Science says . . . ," a more temperate view of the authority of science is necessary.

> I am quite certain that the mass of men believe that the better world of tomorrow will come through science. I think that the belief ought to be publicly combated. Science can indeed produce some of the knowledge and a great many artifacts that will lead to the better life. . . . This alone does not guarantee that the lot of man will be improved by this knowledge and power; indeed,

it is not evident at this point that the knowledge will not destroy man. I think that the point that science alone will not create the good life should be endlessly explored by the press.

The truly remarkable triumphs of science should be presented temperately. Not every clever new device is a historic breakthrough. No new result in science is achieved independently of the tradition and the prior knowledge of science. Perhaps the greatest statement of science is that the universe behaves in an orderly and even a predictable way. Isolated facts are almost meaningless in the sense in which they do not display the essential encompassing character of science. . . .

A perceptive presentation by the press of the interdependence of science and society, the cultivation of the awareness that nothing that science does fails to interact with society and of the awareness that there is nothing that happens in our society that does not have roots in science — these are things that the press must work at if it is to aid in the resolution of a crucial issue of the age, the attuning of man to a science–conditioned world.

Several years ago, Dr. Dael Wolfle, AAAS executive officer, made another plea for doing something more than just dumping isolated tidbits from the research laboratory, medical hospital, or engineering testing ground onto papers. He said interpretation was a special requirement in science writing. In an analogy that newsmen would consider good reporting themselves, Wolfle wrote in the *NASW Newsletter* (December, 1960):

> Compare a science news story with the stock market quotations; if you don't already know what a three point rise in XYZ common means, you get no help from the high, low, close, and net change columns.

Much of the trouble with science journalism in the mass media, Wolfle explained, was with the audience. The average individual does not know enough about science to let the reporter do his best job, he claimed. Then Wolfle continued:

> Much science reporting necessarily consists of individual items. It is the nature of the press to emphasize spot news and it is a characteristic of the way science news is released that it comes out in individual stories. So, Monday's story is on one topic, Tuesday's on another, and Wednesday's on something else again. What these discrete stories add up to depends largely on what the reader can contribute. How good is his background of understanding, interpreting, and fitting together the individual science stories he reads? If he has a good background, the individual pieces fall into place and increase his total store of knowledge and his understanding of scientific activity. If his background is poor, each story remains an isolated item and most of them are soon forgotten.

A few scientists and physicians and some science journalists believe that in the near future there may be room in the mass media for science commentators comparable, if one is an optimist, to Walter Lippmann or James Reston in politics, "Red" Smith or the late Stanley Woodward in sports. If and when this development takes place, then a place will be available for relating the findings of science, technology, and medicine to the everyday living of the average individual. But until that day arrives, science writers will have to interpret, and interpret, and interpret. Sunday science pages, as cited in Chapter 2, offer hope of better going just around the curve.

Dr. J. Allen Hynek, director of the Dearborn Observatory, Northwestern University, and, in the early days of satellites, associate director of the Astrophysical Observatory, Smithsonian Institution, Cambridge, Massachusetts, has had more contacts with reporters than most of his colleagues. So the following suggestions from him carry special validity:

The public reads a great deal about baseball teams going into training in Florida, and paragraphs about whether the manager is going to put this chap in the line-up or that chap, and whether this fellow will be able to play because he has a sprained ankle, etc., but in science reporting the public always seems to want finished results, and spectacular results at that.

It is as though no baseball news were published until the time of the World Series, when things got exciting enough. Why can't the public know about where the various scientific "teams" are "playing," what problems they are up against, etc.?

What are some of the problems? One is certainly the attitude of one's scientific colleagues if one does develop a rapport with the press; he is then accused of publicity seeking. Another problem: the scientist who wants public support yet virtually closes the doors of his laboratory to qualified newsmen. He does this partly because he doesn't want to be bothered but also because he fears censure on the part of his colleagues. It might be good to discuss the whole problem of scientific literacy.

Scientists are becoming aware of the complexities of popularizing their own works, and sympathetic to those responsible for the popularizing. In the early 1960's when I ran three two-week "briefing" seminars for science writers under grants from the National Science Foundation, those scientists I wrote were willing and co-operative when asked to speak to the groups. (Occasionally, as I was lining up both science-writer participants and scientist-engineer-physician speakers, I could not help wondering if one of the more serious bottlenecks in upgrading science reporting might not be news executives and publishers rather than the scientists or the reporters. This reflected a momentary frustration, I now know; in the long run, the media directors released their personnel in superb fashion.)

In defense of communicating scientific facts by means of shared experiences, or still better, through mutual hopes, Dr.

René Dubos, Rockefeller University professor and author, wrote in the "Science and Culture" issue of *Daedalus* (Winter, 1965):

> The breakdown in communication [between the humanists and the scientists of our culture] is complete only when the concepts cannot be related to human experience. The physicist, the biologist, the humanist, and the layman can all find a common ground for discourse if they talk about matter, life, or man as perceived by the senses, or as apprehended in the form of images, analogies, and responses. But discussions of matter in terms of mathematical symbolism, or of life and man in terms of disintegrated components, cannot be related to any form of direct experience. Specialists must return to the original human basis of their work if they want to converse with mankind.

And Dubos undoubtedly would include the science writers among those who should return to "the original human basis" as they go around telling their stories of what is happening on the frontiers of knowledge.

As pointed out in Chapter 3, it is over matters of journalistic format that a real clash of the "two cultures" occurs. And it is here that the knowledgeable scientists, engineers, and physicians often make their most telling thrusts against the contemporary reporting of science in the popular media.

But the truth, demonstrated over and over again in newsrooms across the nation, is that a science story has little chance of being printed or aired if it is not both clear and interesting.

Dr. Eugene Rabinowitch, editor of the *Bulletin of the Atomic Scientists,* made the scientist's typical argument when he said in 1957:

> Good science reporting is impossible as long as its purpose is assumed to be entertainment and not educa-

tion. They [the stories] cannot be only what people want to hear; they often must be what they ought to hear.

John Troan, then science writer for the Scripps–Howard Newspaper Alliance, quoted Rabinowitch's statement and then added this in an article in *Science* (April, 22, 1960):

> Mere publication of a science news story is no guarantee it will educate anyone or make the general public more literate about science. To do any good, a story must capture the reader's interest and sustain that interest. In other words, it must be interesting as well as informative — entertaining as well as educational.

Journalistic formats, many of them traditionally geared to giving accounts of the day's or week's happenings, may be adequate for narrating events of violence, accounts of accidents and disasters, tales of battles, quotations from the market places, and the parade of public meetings. For sports, market quotations, and national politics, the prelude almost always has been reported, and those who "consume" already carry with them the necessary background. But science and, to some lesser degree, medicine — plus a few other news areas such as international affairs — have to be packaged with their own interpretations built-in like the vocabulary glossaries of foreign language textbooks. To many journalists, this straining of conventional formats is almost too much to accept; to most scientists and physicians, such adaptation of format to content seems both inevitable and indispensable.

Journalistic homage to the three tyrannies — deadlines, headlines, and space — *is* rigid, and up to now at least, newsmen have paid that homage. Some newsmen will concede the hazards and danger and will admit that they do not make much sense to a non–journalist, but when neither journalist nor scientist can offer valid, workable substitutes, neither can do anything but shrug his shoulders in frustration.

Dr. Joshua Lederberg, Nobel Laureate at Stanford University, is one scientist who has objected to deadline pressures on the grounds that they too often cause general "hack jobs."

In an oversimplified way, the matter of deadlines may be one of "speed vs. accuracy." But, when a scientist or physician says, as some have, "It seems as if satisfying deadlines were more important than obtaining accurate information," the newsman might reply, "The newspaper gears its life to the deadline: it has to put the papers on the train for future delivery." * And of course the deadline pressures of the electronic media may be even more exacting than those in the print media.

And so, despite the accumulating evidence that deadline homage may possibly be rooted more firmly in tradition than in logic — and despite the recurring question, "Couldn't a carefully prepared second day story better serve the reader and listener than a hastily and perhaps sloppily prepared item done just before the deadline?" — the deadline ritual seems intrenched for some time to come.

Another problem: headlines on science, technical, and medical stories may stimulate some individuals to ask, "Who writes those damned headlines, anyway?" The straight-faced answer would be, "A man on the copy desk." On most newspapers, these copy editors possess a wide variety of talents and skills, but training in science and medicine may not be one of them. A scattering of metropolitan newspapers now have copy editors with special background who edit this kind of news item, but on the rest of the nation's dailies it is catch–as–catch–can. Until the ratio of science and medical news comes closer to that of sports, business and financial, society, or local news, it is unlikely that most newspapers and smaller radio and television stations will have a science and medical news specialist at the copy desk; certainly it will be long after they have named a part-time science news writer, which many of them have not yet done.

Some years ago, Louis Lyons, then curator of the Nieman

* See "Why are reporters always in such a hurry to print a story?" in Krieghbaum, *When Doctors Meet Reporters*, pp. 54–61.

Fellowships at Harvard, illustrated how the same science story could receive headlines that carried vastly different impressions.* The same story about an Atomic Energy Commission report received these headlines:

A-TEST TOLL TO BE SMALL, AEC SAYS
— *The Pittsburgh Press*

RADIOACTIVE CARBON TO KILL
172 CHILDREN A YEAR
— *The Cincinnati Post*

More recently, Dr. Albert Gilman, chairman of Pharmacology at Albert Einstein College of Medicine, Yeshiva University, collected these headlines that appeared over a three-paragraph Associated Press report on a two-month-old study of monkeys who were being taught to smoke in an effort to find out why people like cigarettes:

Study Monkeys to Learn Secret of Smoking
Monkeys Taught to Smoke in Research into Habit
Monkeying Around in Cigarette Study
Monkey See Monkey Do

If science and medicine need more interpretations than most other kinds of news, then more news space must be earmarked for the expanded articles. Since the total amount of the "news hole," as it is sometimes called, remains fairly constant from week to week, large allocations to science and medicine must be made at the expense of other news. This means missionary work with editors and publishers — and the fate of missionaries in a hostile country is not always a happy one.

A revealing study of attitudes held by scientists, science writers, senior news executives, and readers by Dr. Kenneth G.

* See "The Bomb and the Press" (No. 3 in a series of television programs produced by WGBH-TV, Boston, Mass., with a grant from the Fund for the Republic), p. 8.

Johnson in the *Journalism Quarterly* (Summer, 1963) showed that editors used substantially the same criteria for judging science news as the non-science readers use, not those of the scientists, science writers, or science readers.

Scientists may not be as difficult to please as some might have us believe. It is repeatedly said that these specialists want the facts presented in a writing style that is is impersonal, passive, and straight-forward, without the embellishment of interest-arousing devices. However, Dr. Norman McKown reported results that run counter to the accepted tradition in *Studies of Innovation and of Communication to the Public* (Institute for Communication Research, Stanford University, 1962). He sought to introduce personal pronouns and an informal, sometimes humorous, style into a 15-chapter document of 172 pages; he then tested the manuscript on an audience of scientists, receiving 98 questionnaire replies from scientists and scientist-administrators and 43 from military project personnel. How McKown did this is shown by this quotation from the opening chapter:

> You have a project, you have the green light, and you have just been received into that stalwart fraternity who march resolutely to the blistering sands of Nevada and the soggy atolls of the Marshalls, on alternate years, to measure the effects of nuclear weapons. If this is your first field junket you'll find conditions very different from the laboratory. If you've been on field tests before, but never on nuclear field tests, you'll find a number of peculiarities. If you've been on nuclear field tests before you can stop here. You are probably only too well aware of these peculiarities.

Another example from a section on meteorological conditions began:

> There are five seasons in the Marshalls, as there are in Nevada: fall, winter, spring, summer and test. Test sea-

son generally comes in late spring and early summer because of the stability of the climate at that time. (This is what they say, but everyone knows that although we've harnessed the atom no one has saddled the weather.)

Seventy per cent said the use of personal pronouns was not excessive and another 23 per cent replied, "Not particularly"; only 6 per cent checked "Definitely yes" or "Yes." Even more (83 per cent) said their use detracted "not at all" from the total effectiveness of the manual and 14 per cent more felt it subtracted "not very much." This meant that 97 per cent found personal pronouns to be no noticeable source of irritation.

Responses to the question of informal style followed roughly the same pattern. Sixty-eight per cent replied that informality was not excessive, 23 per cent took a mildly favorable position and 8 per cent said informality was excessive. Eighty per cent said the informality detracted "not at all" from the total effectiveness and an additional 13 per cent felt informality was not a very serious distraction.

Scientists, engineers, and physicians frequently ask why science correspondents do not bring their copy back to them to insure the accuracy of the report. Some newsmen do — while others may have their research assistants doublecheck on names, figures, and quotations, almost as part of the routine editing process. But most of those who cover science for the mass media do not go as far as showing the actual story or script to the source which supplied information. First, with the pressure of deadlines — and there is always a period when time is short, even if it comes only once a month — writers do not have the minutes or hours to telephone to ask one more time if the way they phrased their interpretation squares with the thinking of the news source. Second, newsmen often find that scientists confronted with a piece of copy, even one where the facts are correct, cannot resist the temptation to turn into editors, which is neither their profession nor their responsibility. Inaccuracies, certainly, should be pointed out, but that is all. Too often a news source pushes commas and phrases around instead of simply insuring that the facts march with precision. This has happened so often in the past that many newsmen reject the whole idea of showing their copy to anyone other than their staff associates. Sometimes this feeling veers close to viewing such changes as an attack on the reporter's professional competence and integrity.

Legitimate complaints have been voiced about the lack of training of some writers assigned to cover science stories.

Some newsmen, as witness the information given in Chapter 6 on college courses the science correspondents had taken, do not have sufficient preparation for the work they are now doing. But even more serious for the mass media representatives is the problem of keeping up with those who are far out on the frontiers of knowledge discovering new facts, practices, and concepts. Even the best possible training a decade or two ago would be obsolete for covering what are the very latest achievements.

Dr. Harold C. Urey, whose work on "heavy water" won him the 1934 Nobel prize in chemistry, reported that in many cases he found a lack of elementary knowledge of science on the part of reporters, adding, "Often the newspapers reporting stories confuse such elementary terms as energy, force, acceleration, etc." He also said there seemed to be a "persistent confusion of science and engineering," although it was perfectly true that scientists sometimes did things on engineering subjects and that engineers sometimes did things which deal with science.

When he presented the first American Medical Association awards in 1965 for "outstanding examples of journalism that contribute to a better public understanding of medicine and health in the United States," Dr. Wesley Hall, vice-chairman of the AMA Board of Trustees, said:

> All over the nation writers have developed who are specialists in reporting medicine in the newspapers and magazines and on the air. These writers have learned they can turn to the medical profession for help and

guidance in interpreting and translating medicine to the public. The writers themselves have studied and learned much about medicine in an effort to do a better job.

Dr. Glenn T. Seaborg, chairman of the U.S. Atomic Energy Commission and a Nobel Laureate himself, probably spoke for a majority of his colleagues in science when he told the 1963 national convention of Sigma Delta Chi, professional journalism society:

> The trained science writer no longer is a rarity, although I believe his ranks still need to be increased.

But there is another aspect in this argument about the science journalists' training, as Earl J. Johnson, former vice-president and editor of the United Press International, stressed in one of his 1960 weekly letters to telegraph editors:

> A science writer who comes to identify himself with scientists in his thinking and writing will soon lose his non-scientific readers. A science writer must be first and always a reporter and writer. The ability to make a topic come alive in print is an art. It requires imagination with words, constant practice and self-discipline.

And so does covering science for radio and television.

11

BARRIERS
Internal

No matter how much they may disagree on other aspects of the popularization of science, technology, and medicine in the mass media, those who write the news and those who are the sources for such information agree that the reporter's assignment must be complicated, difficult, and exacting if it is ever to approach success.

Dr. J. Robert Oppenheimer, the physicist who helped to develop the first atomic bomb and who later became head of the Institute for Advanced Study at Princeton, undoubtedly spoke for many scientists when he said some years ago:

Nearly everything that is now known was not in any book when most of us went to school; we cannot know it unless we have picked it up since. This in itself presents a problem of communication that is nightmarishly formidable.

Most science journalists would agree that their job is "nightmarishly formidable" at times. Yet many authors believe that it is possible to popularize science news and information successfully, despite the difficulties. For instance, Martin Green* has written that the science writer should avoid condescension and think of his lay reader as "someone as intelligent as himself in non-scientific matters, someone with as much taste, tact, and general education as himself." Translating technical terms into every-day language must follow the same rules as any other kind of good writing, or, as Green put it, the newsman "must imagine a reader like himself, only more intelligent, and with plenty of adult experience, but ignorant of this particular subject." Intelligent laymen need this approach in nine-tenths or more of the subjects they try to understand. Thus, science reporting becomes as keen a challenge as any in the contemporary age of mass journalism.

Barriers to effective popularization of science, technology, and medicine can be either internal — intrinsic to the nature of the assignment itself — or external — imposed from the outside by non-journalistic pressures.

Probably primary among the internal barriers is the tricky and difficult job of translating complicated and technical details from the language of science into the dialect of the average individual. But right along with this is the need to generate some degree of scientific sophistication so that readers and listeners are able to appreciate the concepts behind the real meanings of the words and phrases being used.

* See Martin Green, *Science and the Shabby Curate of Poetry: Essays about the Two Cultures* (New York: Norton, 1965). Note especially his sections on "Popularization" (pp. 32–41) and "Science non-fiction" (pp. 139–55).

Difficulties of translating into the vernacular of popular journalism have been discussed previously as part of the clash between the two cultures of science and communications and again as one of the problems facing reporters. They are obviously part of these aspects of science reporting, but in this chapter we will look at them strictly as barriers to meeting the goals of effective transmission of facts, ideas, and concepts; there will also be some discussion of what might possibly be done about lessening these impediments.

Despite the flexibility of English (and it is probably better adapted to science than any other language), the scientists have been forced to invent a better vocabulary for scientific purposes. It is from this richer and more precise language that the science writers have to translate.

M. W. Thistle, a scientist who became chief of the Public Relations Office of the National Research Council of Canada, wrote about it in an article reprinted in *Science* (April 25, 1958):

> To ask a man to translate from one or several rich, relatively new, and precise scientific languages into a single poverty-stricken language of inadequate structure — with the built-in faulty science and outmoded thinking of previous centuries showing at every seam — is asking a very great deal. Whatever detail this man does manage to get across to a general audience will certainly be distorted and, to some extent, actually false. No other outcome is possible.
>
> If you listen closely to a man who is supposed to be good at talking about science in English, you will notice that he is not trying to transmit many details — he is giving broad outlines, general trends, and a highly condensed abstract of results. If he really is good, he knows that scientific details cannot be transmitted in an undamaged condition to a lay audience, or even to a nonspecialist scientific audience.

Yet Thistle felt that both scientists and journalists should work hard to push the science message across to the general public. And he gave this concrete advise in his final paragraph:

> In conclusion, I might point out that we have some very old precedents for breaking through the barriers and talking to ordinary folk about extraordinary things. Jesus had such a problem. His technique was to put what he had to say into a perfect little short story, dealing only with familiar things that you can touch and see. He would begin with, "A certain man had two sons," or "Behold, a sower went forth to sow." To this day, the only device I know that will actually work for an audience of fishermen, tax-gatherers, publicans, housewives, or other groups of laymen is this same technique of analogy, comparison, metaphor, simile, and parable.

Some metropolitan dailies' editors, however, believe that their reporters do not have to use this approach and that science stories in their general-readership publications need be understood only by their scientist-readers, who in these few newspapers comprise more than the usual percentage. These executives feel that such articles build prestige in the scientific community. But, as one respected science writer said, "It makes for a dull newspaper." Then he added that he would not write any science or medical story unless he at least thought he understood it. "That goes for deoxyribonucleic acids and the crumbling laws of parity." Both were topics on which he had written very little.

The controversy over translating science news boiled at its liveliest during the mid-winter, 1964, meeting of the National Association of Science Writers after a talk by Dr. Percy H. Tannenbaum, director of the Mass Communications Research Center at the University of Wisconsin. He said in part:

> Our findings have indicated a relatively strange situation — at least for a mass media type of operation — in

that a lot of the effort which goes into the selection, preparation and presentation of science news to the public
seems directed more at the non-reader than the reader of
science. Our research has demonstrated, for example, that,
in TV at least, the professional communicators' estimation
of the public beliefs about mental health and illness is
quite different from what the public actually believes. In
the newspaper area, we have found that the desk editor
— the so-called "gatekeeper" who is responsible for what
gets into the paper and how it is treated — uses a set of
criteria in judging the value of science news stories that
are quite at odds not only with the criteria set by the
professional scientists, but by the science writer and,
most important, by the reader of science news. The way
the editor makes his judgments in this vital area of news
appears to be most closely related to the way the non-
reader makes his judgments. This state of affairs also is
reflected in our research dealing with the specialized language of science, and attempts to "translate" scientific
terminology back into lay language. All too often, this
leads not only to an alienation of the scientists but, more
critically, of the self-selected part of the audience most
interested in science news qua science news.

My suggestion, then, is that we recognize that only a
part of the total public is interested enough in science
to want to select and attend to science information, and
that we direct our attention to that select audience,
rather than the vast audience of uninterested people.
The professional science writer appears from our research to be quite admirably suited to the important role
of mediator between the scientist, as news source, and
the science reading part of the total audience. That more
and better such specially-trained science writers are
needed is obvious. But perhaps more important is the
fact that they must be allowed to exercise increasing
control over the selection and preparation of specialized
science news for the specialized audience. They, much

more than the general editor, are more "in tune," as it were, with the audience — just as the sports editor is in tune with his, the society editor with hers, the business editor, etc. It should be recognized that the newspaper particularly, and the other media generally, rarely if ever cater to the interests of a complete or near-complete audience with any one given story. Rather, they present specialized content in a specialized style for specialized sub-publics. Science communication is another such specialized communication situation, and it would be of advantage to all parties concerned — the scientists, the media, and the audience — if it were recognized as such.

Martin Mann, then NASW president, picked up the argument with these remarks in the Association's *Newsletter* (March, 1965):

> It is true that most of us [science writers] write science articles for "nonreaders." We try to make our stuff understandable and (hopefully) appealing to a mythical "average American." To the extent that all average Americans do not read all science stories, we write for nonreaders.
>
> This seemingly inefficient approach is logical because the "readers" and "nonreaders" won't stand still. Neither group is well defined and static but fluctuates rapidly, wildly, and erratically. Before 1957 almost all average Americans were nonreaders of stories on space rocketry; after Sputnik nearly all average Americans became readers of this kind of story. Similar conversions of nonreaders to readers (and back), less drastic and also less predictable, go on all the time. Who is a reader, who is a nonreader depends on the story, the time, politics, weather . . . So it seems proper to aim any story that could appeal to a broad audience at a *very* broad audience: write for the nonreaders.
>
> This is why Professor Tannenbaum's sports page analogy is misleading. You can't compare science news to

news of such major sports as football and baseball, which are familiar to nearly all Americans. You could compare science news to news of a minor sport such as hockey (Montreal and Boston readers will please forgive me!). It is logical to write hockey news solely for hockey fans; their number will not fluctuate erratically. Developments in hockey are unlikely to be connected to public health, private business, or national prestige, so there is little opportunity for the nonreader–reader conversion, so important to science news, to affect hockey news. No great harm is done, few potential readers are lost, if only true-blue hockey fans can understand the hockey stories.

To write science news that way is impractical anyway. If we limit our audience to those who already "want to know," some other writers and editors will figure out ways to attract the larger audience: those who don't yet know they want to know.

Regardless of whether they agree with Tannenbaum or with Mann, reporters are almost always sure to discuss how to translate science news when they get together for bull sessions after finishing their day's assignments at the formal convention program. Sometimes cited are the 1957 and 1958 NASW-NYU surveys which uncovered evidence that a more humanized approach to science reporting (or as the journalists would say, a more readable version) did not turn away intelligent readers. A series of paired choices of topics showed that all groups, including the college educated and those who had taken courses in science, preferred the more concrete or humanized language to the stereotyped, abstract labels.

Blair Justice of *The Houston Post* reported this observation about the readers in his community, which is not unlike other expanding localities:

> Although the Houston area has a great number of scientists, and even more technologists and engineers, we

do not find that they skip the science page simply because it is in the language of the layman. Some even say that, with science being so vertically oriented, our page helps a scientist working on tranaminase enzymes to learn something about quasars and other out-of-field subjects.

In their translating jobs, science writers may carry into their own writing, probably almost unconsciously, some of the textual qualities of the reports and papers they read and study. Dr. Wayne A. Danielson, dean of the University of North Carolina School of Journalism, analysed the Associated Press wire report for the week of September 28–October 4, 1964, with the aid of a computer. He and his associates found that the sentence lengths of news items on science, invention, and research were the highest of any major categories and more than two words a sentence above the over-all average of 18.7 words.

While Danielson wisely pointed out, "It is ridiculous to write to please a formula — because formulas are not measures of *good* style, merely of *difficulty* of style," the haunting doubt remains that science journalists may be losing some segments of their potential popular audience without really realizing they are writing over their readers' heads. If headlines on science stories consistently signal hard reading to come, may not some readers learn to skip over these news items?

No matter how well the translating job is done, to some extent the science journalist is dependent on how much sophistication the reader brings with him. In the days of primitive realism, an individual could blame his failure to understand on the whim of a god — or of a devil. Then came the beginnings of science with the development of rulers, thermometers, telescopes, and microscopes and the resulting achievements of the classical scientists such as Galileo and Newton. Now another stage, usually called modern science, is based on the work of such individuals as Albert Einstein and Niels Bohr. Those on the advancing frontiers of basic research may be pushing into yet another phase, but they are so far ahead of most of the public that even their fundamental concepts often appear all but impossible to translate.

The popularity of much folklore in the mass media, even in the 1960's, indicates that a large number of individuals (some of them working in communications) have hardly begun to venture away from the initial stage of primitive realism. For instance, consider the coverage given to Groundhog Day, Friday-the-13th traditions, black cats, horseshoes, rabbits' feet, and all the variety of ghost stories, witchcraft, and "psychic phenomena." As recently as 1956, one New York City metropolitan daily devoted almost three columns of its front page to a Groundhog Day story and its accompanying picture. Woolly bears as prognosticators of the kind of winter we will have were far more popular for some years than anything that the U.S. Weather bureau could compile — or even the *Farmer's Almanac*. *The New York Times* in 1966 quoted "one professional astrologer" on the qualities of "the average Pisces" and implied that they were true for Lynda Bird Johnson, daughter of President and Mrs. Lyndon B. Johnson, who "comes under the 12th sign of the zodiac by virtue of her birth in 1944 on March 19."

Infrequently even the science writers' bosses may be a barrier to intelligent reporting because they lack the sophistication to make meaningful assignments. One veteran who had been writing about science for decades told of an experience several years ago when he was asked to write a feature on a new operation for extracting salt from the waters of Lake Erie.

"I don't know how that could be done, for Lake Erie is fresh water," he told his editor.

"Well, if you don't know, why don't you try to find out?" replied his boss, entirely sober-faced.

"Working conditions," to borrow a general phrase from the language of the labor negotiating table, cannot help but play some influence on the quality of science news coverage. The battle against the deadline, the summary lead paragraph giving the "guts" of the news in 50 or so words, the inverted pyramid organization of conventional news stories, the brief broadcast condensation, and the editing and headlining of science stories on many newspapers by non-science-oriented copyreaders — all these have been cited as barriers to effective science communication by the mass media. Too often the critics have been right. All of these points have been mentioned earlier, but let us take another look, this time from the point of view that they pose obstacles to good and effective reporting, and see if something might not be done about them.

All of the mass media representatives are possessed with a passion to give the news when it happens — or as soon as possible thereafter. It is certainly a tradition for newspapers, news magazines, radio, and television despite the claim, usually of non-journalists, that this idea is an antiquated cliché. Newsmen treat events as if they were milk or fish standing in the noonday sun without refrigeration: an article of highly perishable quality. In many ways, the news staffs and their superiors are right. No one wants to hear or read the same old news. Yet radio and television news flashes can beat the most modern daily's presses and the newspapers may be days ahead of the weekly news magazines. These are conditions that all media have learned to live with.

Is there a lesson in adjustment that might be applied to science news generally?

Newspapers meet radio and television competition by supplying background, explanation, and interpretation which often are crowded out of the usually brief broadcasts. In other words, the dailies back up spot-flash information with the facts in perspective. News magazines, with a week's time to organize the world's events, often can synthesize a series of happenings so that they make more sense, can better show cause and effect relationships, and can focus the material to give readers a sharper picture. Sunday newspapers, if they wish — which all of them do not — can perform the same functions as the news magazines. Thus, the newness of the news is maintained by bringing in more information to yield better perspectives.

Skilled public relations approaches, too, may enable reporters with even the shortest deadlines to hurdle the disadvantage with ease and to supply effective stories. Advance news releases or pre-release briefings provide the needed time that often makes the difference between a story pushed through in machine-gun bursts of speed and one written in the more leisurely (and probably more accurate) "fire at will" style of heavy artillery. To illustrate, let me cite the differences between the University of Michigan's April 12, 1955, release of the Francis report on the effectiveness of Salk's polio vaccine and the U.S. Public Health Service's January 11, 1964, release on the study of the relationship of cigarette smoking and cancer.

At Ann Arbor, correspondents scrambled to grab releases on a truck being wheeled into the auditorium and then rushed to telephones with only the slightest chance to do more than glance over the summary abstract before they began dictating their initial news stories. Most of the more serious science writers at Ann Arbor felt there should have been some time to look over the text, charts, and tables before they had to start rolling the story out to the American public. One reporter summarized the reaction, "Most of us really got the sense of it that night, going home." Others said that fortunately for them and for the public the initial impressions did not have to be modified or reversed after further study.

When the cigarette smoking and cancer report was released in 1964, the U.S. Public Health Service assembled approximately 250 newsmen in time for them to have several hours to study the 387-page report and thirty minutes to hold a news conference. During this period, the doors were locked; then the correspondents were released to telephone their stories. The Public Health Service had bridged the time-deadline barrier by allowing hours for studious examination of the report itself plus a news briefing to clear up any ambiguities in reporters' minds. It had done all this before the scheduled deadline when the scramble to be "firstest" would take place.

When the Sixth International Congress of Biochemistry met

in New York City during the summer of 1964, Dr. E. H. Ahrens, Jr., of Rockefeller University, and then chairman of the Press Committee, got his colleagues to prepare a 115-page booklet, *Concepts in Biochemistry*, with essays covering the ten areas of interest into which the 208 symposium papers and 480 shorter scientific reports would logically fall. The objective, Ahrens explained, was "to present concepts rather than facts alone, and to do this in sufficient depth for science writers to gain a solid grasp of the background of the subject." Originally 500 copies were printed and distributed well in advance of the Congress so that newsmen could prepare themselves by reading the booklet. The response of biology teachers, physicians, public health officials, and college students was so enthusiastic that the Organizing Committee of the Congress financed a second printing of 12,000 copies "to spread a deeper understanding of biochemistry to as many interested readers as is financially feasible."

In a less thorough fashion, other organizations frequently include a "background" release along with the spot developments they are giving to the news media. If impeccably honest, these "backgrounders" can be most useful to science writers, but the temptation to do a little card stacking in the client's interests is too great for some individuals to resist.

Although the summary lead paragraph and the inverted pyramid organization are two ancient traditions of newspaper journalism with much to be said in their favor, both can deaden reader interest in any story. If an appeal is to be made to the "nonreaders" mentioned by Tannenbaum and Mann, then these two stultifying approaches should not be allowed to drive away readers of any sort. This means that there is a great need for science reporters to attract attention with a bright and sprightly lead and an intriguing organization. Otherwise, the conventional nonreader of science news continues in that category.

On a relatively short news broadcast, radio and television compress the entire science activity into little more than the equivalent of the newspaper's summary lead — with or without film clips. Even pictures, when the time is so short, serve as little more than "alerters" in news communication, just as the bob-tailed story does in a newspaper.

Newspapermen are breaking out of these journalistic straitjackets more often, especially when they have more space in, say, a Sunday edition. And broadcasters seem to be turning increasingly to documentaries dealing with science, technology, and medicine. In all these cases, a sympathetic nod is needed from news executives in order for the working newsman to get more space and more time.

Typographical errors are apparently inevitable booby traps for journalists and they plague science writers just as they do everyone else. One of the more widely publicized examples happened years ago when the late Howard W. Blakeslee was covering an American Medical Association session about cardiovascular disease. The Associated Press science writer began a sentence with "These arteries . . ." but he struck an extra letter at the end of the word "these" and crossed it out with an "x." A baffled AP telegraph operator, unable to find Blakeslee and check, sent the wording, "The sex arteries . . . ," and that was the way it was printed in many newspapers.

Possibly more than in other news fields — politics, sports, or business — headlines on science stories bother the people whose names are mentioned. Others may react in the tradition of the individual who exclaimed, "I don't care what you say about me as long as you spell my name correctly." Not so with most scientists.

Much of this criticism of headlines rests on a blissful ignorance of the inexorable inelasticity of type. United States newspaper headlines have to fit an exacting measure of so many — and absolutely no more — letter-units per line. Headline writers know this, only too well; scientists do not. Since this information is not part of their daily needs, there is little reason why they should know about it — until they start to comment on what they consider faulty headline composition. Some journalists believe that not even modern headline reforms manage a truly logical approach; they may be correct, but newspapers will follow the current rules until a major headline revolution comes along.

A few metropolitan dailies have science specialists to handle such news copy. This trend probably will continue and possibly will expand but, until science news receives as much space in papers as sports or society, it is unlikely that the economics of publishing will permit much acceleration in the hiring of science-oriented copyreaders.

Popular mores, inside and outside the news room, influence the ways in which an editor assigns newsmen to cover events — and the ways in which newsmen approach their sources. Physicians have been with us for several thousand years and have dealt directly with the public, which has become acquainted with the benefits. Thus, medicine became established as a familiar, and generally accepted, profession. Modern science, in one commonly used sense, has been around for approximatly 300 years and, in that time, it has gained a special stature, being admired and respected for its cargoes of better standards and longer life, although it often failed to enjoy public comprehension. This was due, in part, to the degree of specialized terminology and abstract concepts which required translation before they could be understood, even in vague sense, by the typical man in the street. The social sciences, largely products of the twentieth century, have advanced on everyday social problems, but these disciplines are so recent that people who have stereotyped images of social activities have not yet had time to modify their concepts to correspond with the new realities. Many members of the general public and some newsmen stll consider themselves their own best social scientists.

"I know how kids should be dealt with," is a fairly common remark. "Who needs a social scientist to tell when to spank a child? Or what a teen-ager should learn in high school?"

"I've got the answer to how the slums could be cleaned up. Why listen to those fancy experts?"

If there is a major expansion in science reporting in the decade or two ahead, much of this should take place in improved coverage of our social problems and the scientific attempts to overcome them.

A sign that this trend is already underway was the presentation of the 1965 Sigma Delta Chi Award for Distinguished Service in Journalism for General Reporting to Alton Blakeslee of the Associated Press for a 10-part series on the effect of science and technology on the individual. Judges for the professional journalism society called it "an example of a new dimension in science writing — an expansion into social science reporting."

Yet many newsmen would agree with Murray Kempton, newspaper and magazine columnist, when he called sociology a "remorseless pursuit of proof of what everyone knew all along."

The mass media, which need to be triggered by events before they can supply essential background, too often by-pass the massive, slow movements in society. It is like a cavemen's gazette failing to cover the coming of the Ice Age and, instead, concentrating on reports of dances around dimly lit campfires. For most communications media, it took a Hiroshima to send their reporters to the $E=MC^2$ formula that had been underlying it all for three decades. The novel innovation, regardless of how important, may be submerged by the banal, current spectacle.

We do not have to abandon hope, however. As E. G. Sherburne, Jr., head of Science Service and former director of the AAAS's program for public understanding of science, said there is no real reason to believe that popularization, with the proper efforts, is not possible "on a very high qualitative level, acceptable to the communicator as artistic and dramatic, and to the scientist as honest and accurate." And, he added, when the right kind of communications is available to the public, there will be no difficulty in obtaining either audiences or understanding.

12

BARRIERS
External

Science writers face a wide variety of handicaps to effective and honest reporting due to outside forces and pressures beyond the newsroom barriers and the very nature of communication itself. These comprise: governmental and corporate censorship, authoritarian and over-zealous bureaucracy, and the pleadings of special-interest groups.

One of the constant struggles is between those who think they need (or deserve) secrecy and the newsmen who have a duty and responsibility to find out and to inform the public of a democratic society. This is true for science writers as well as journalists assigned to cover any other news. The competition may be just as keen to ascertain the site of a new federally-financed accelerator as it is to find out the location of a new dam or federal building, to learn about a new advance in the detection of cancer as it is to uncover the route of a presidential inspection tour into the newsman's home territory, to describe a new elementary particle as it is to disclose the latest Hollywood divorce petition.

In the immediate post-World War II period, the sharpest clash was between science, which rests on a free exchange of ideas and discoveries, and secrecy, which seeks to protect the national security. They are often a pair of true opposites and thus in conflict. By post-war statute, the Atomic Energy Commission was equipped with comprehensive control over atomic energy information and these laws defined the term "atomic secret" in a special way: "All data concerning the manufacture or utilization of atomic weapons, the production of fissionable material or the use of fissionable material in the production of power, but shall not include any data which the Commission from time to time determines may be published without adversely affecting the common defense and security." Thus, the act provided that any fact or document in the area of atomic information, no matter whether it already had been published or had otherwise become well known, was declared a "secret," unless and until the AEC decided that it might be published without adversely affecting the United States common defense and security.

One well-publicized confrontation over the application and ramifications of the "atomic secrets" act took place when *Scientific American* proposed to print an article by Dr. Hans A. Bethe about the then-projected hydrogen bomb and the Commission obtained deletions in the magazine's issue for April, 1950.

In mid-March that year, the AEC announced a new extension of its censorship policies in telegrams to its own staff members and consultants, and to employees of its contractors. The Commission said it wanted "to avoid the release of technical information which even though itself unclassified may be interpreted by virtue of the [hydrogen bomb] Project connection of the speaker as reflecting upon the Commission's program with respect to thero-nuclear weapons."

Gerald Piel, publisher of *Scientific American,* said the effect of the telegrams was to inform scientists connected with the hydrogen bomb's development "that they could teach physics, but that they cannot talk in public about the unclassified, that is non-secret, technical information concerning the hydrogen bomb — which is, of course, the only information they would talk about in public."

Some weeks earlier, Bethe had agreed to write an article on the projected bomb. Before World War II, he had developed the theory of atomic fusion to explain how the sun and other stars generated their light and energy. During World War II, he was chief of the Theoretical Physics Division of the Atomic Weapons Laboratory at Los Alamos; in 1950, he was professor of physics at Cornell University and employed as a contract consultant to AEC. With this background, Bethe was naturally well aware of the propriety required of a contract consultant who had access to "secret" information; he knew it was necessary to take precautions before entering public discussion of the issues surrounding the hydrogen bomb and naturally checked to see that the facts he included in the article already had been made public. As a courtesy, and to keep colleagues informed, Bethe sent copies of his manuscript to a number of associates. As Piel said, this was done "not for review but for their information." One of these copies went to a scientist who was an AEC member.

Piel told what followed in a speech to the American Society of Newspaper Editors on April 21, 1950:

> On March 15th, after the April issue of *Scientific American,* containing Dr. Bethe's article, had gone to press, we received a telegram from the A.E.C. requesting that we withhold from publication the technical portion of the article.
>
> *Scientific American* stopped its presses and asked the

Commission to specify its objections. After it agreed to review the article the Commission's objections came down to a request to delete several sentences. In order to be able to proceed with publication of this issue we complied with the Commission's request.

The Commission then asked that all copies of the original article be destroyed. An A.E.C. security officer visited our printing plant and supervised the destruction of the type and the melting down of the printing plates with the deleted material and the burning of 3000 copies of the magazine which had been printed before the presses were stopped. Thereafter we put the magazine, with the expurgated version of Dr. Bethe's article in it, back on the press and proceeded with the publication of the issue.

Piel said that he believed the "real objection" involved not the statements themselves but the fact that an informed authority was making them. The effect was, he felt, a signal to all scientists concerned to "keep their traps shut," a phrase which actually was used by an AEC commissioner in discussing the case. Yet it is true that what might be an outsider's informed guess might also constitute a disclosure of classified information if made by one who had access to "secrets."

Years later, when the original Bethe article was declassified, the *Scientific American* staff compiled citations of prior publication for the deleted sections, many of them written by Bethe himself during the 1930's, to establish that the initial draft did not, in fact, include unpublished and secret information about atomic energy and its uses in a hydrogen bomb.

Officials do not need special legislation to curtail information. For instance, the National Aeronautics and Space Administration's Manned Spacecraft Center stamped a paper entitled "Functions of the Public Affairs Office" with "For Internal Use Only" in an effort which a Houston newspaperman called "a new kind of classification to evade the rather clear rules defining classified information."

While modern science and technology have widened the
range of potential subjects that might have to be kept undis-
closed in order to maintain a military or diplomatic advantage,
other pressures have pushed just the other way, for publicity
is a cornerstone of both science and democracy. This was explained by Dr. Francis E. Rourke in *Secrecy and Publicity*
(Johns Hopkins Press, 1961):

> Much of the scientists' criticism of secrecy rests on the
> premise that it is unnecessary, because it attempts to con-
> ceal matters that are part of the general fund of scientific
> knowledge in the Western world or that can be easily
> discovered by scientists working outside the classification
> system. There is, of course, general agreement that the
> weapons developed through modern science may them-
> selves need to be kept secret . . .
>
> The difficulty of keeping secrets is related to what is
> perhaps the most basic of all scientific criticisms of gov-
> ernmental secrecy. This is the charge that it has become
> unduly negative in character through its stress upon the
> necessity of concealing rather than uncovering informa-
> tion. In the view of an influential part of the scientific
> community, innovation and development depend essen-
> tially upon a widespread flow of communications among
> scientists working on related problems. Insofar as govern-
> mental secrecy stifles such communication, it hampers
> scientific progress. From this perspective, the benefits to
> be gained in terms of a high rate of scientific discovery
> more than offset whatever element of risk may be present
> in removing restrictions on communications. Scientific
> achievement is itself regarded as the firmest basis for
> national security. . . .
>
> The central fact with which all chief executives must
> now come to terms is that a classification system designed
> in the short run to protect national security may at the
> same time choke off the flow of communications essential
> for scientific development and the long-run interests of

national security. While the union between science and warfare in modern society has made executive secrecy increasingly necessary, it has also made it increasingly hazardous. Presidential statesmanship of a high order may be called for in the future if it proves necessary to buy scientific progress at the price of greater disclosure of scientific information. Reliance must be placed upon executive initiative to take this risk, in view of the fact . . . that Congress, although predominantly inclined to favor greater publicity in wide areas of executive activity, has tended to give uncritical acceptance to the need for secrecy in matters affecting the national security.

Many of the arguments for a governmental policy supported by an informed public opinion apply to the areas of modern technology; they indicate a curtain be lowered over details of applied know-how for military uses, not over the fundamental concepts involved. After all, it is the public that undoubtedly will pay the price in taxes and, if the worst happens, in atomic annihilation. But where secrecy intervenes, an informed public is impossible. Then, as Dr. I. I. Rabi, Columbia University professor and the 1944 Nobel Laureate in physics, pointed out in *The Atlantic Monthly* (August, 1960), "Too often we have, instead, a manipulated public opinion formed by leaks, half-truths, innuendoes, and sometimes by outright distortions of the actual facts."

As public attention moved from atomic weaponry to space exploration, so did the fight to tell the public about what was going on. Only tangentially did the issue veer to national security — this was when military applications were concerned — but more than that, it was an attempt to gain a Cold War advantage by making United States efforts look good in comparison with those of the Soviet Union. On rare occasions, the truth got a thin layer of synthetic gold leaf to make an event seem more spectacular than it really was.

For instance, after the United States launched its talking Atlas satellite in December, 1958, two reputable newspaper sources criticized the way this news was released to the public.

Peter Edson, in a Scripps-Howard newspaper column on December 29, 1958, included this comment:

> No one denies this was a successful and important launching. But in trying to build it up into the greatest thing since the original appearance of the Star of Bethlehem, a lot of claims were made that simply weren't so.

Further, an editorial in *The Washington Post* of December 23, 1958, included these statements:

> It is now pretty clear, however, that the project was principally a publicity stunt. This newspaper sensed deception and declined to be a party to it. But the Administration helped to create the impression elsewhere that the Atlas is the biggest satellite yet launched. This is untrue except in one narrow sense: the second-stage rocket and the payload were designed to remain together. The last Soviet sputnik had instrumentation of greater weight than the Atlas carried and the rocket used to propel it had considerable more thrust. . . .
>
> The worst part of the whole business is that the Administration information policy made the press generally an unwitting accomplice in the propaganda. . . . From all evidence this misinformation was deliberately cultivated.

Sometimes the censor's blue pencil is used blatantly with embarrassing results — for arrogant editors. For example, when William Hines of the *Washington Evening Star* submitted copy for a story to be transmitted by the Army via Courier satellite in 1960, he included a section which read:

> It was a stunt, of course. A gimmick. But there was an aura of history about it that made participation in the stunt impossible to resist.

However, the article, changed by an Army signal corps "technician," was transmitted to read: "This historical experiment may well be the forerunner of tomorrow's mode of communication."

The Army Office of Information, red-faced, admitted that "the authority to change matters of fact appears to have been misapplied" when the whole affair was investigated by the chairman of the House Government Information subcommittee, Rep. John Moss, D., California.

Do news sources generally give reporters the truth, pure and uncontaminated?

"Hell, no," was one nationally-known science writer's reply, and then he added that government agencies, with which he primarily dealt, will lie "at every turn and opportunity." But the science writer has a powerful weapon to wield against the misinforming news source. He can print the lie attributing it to the source by name and then in the following paragraph state the truth as he knows it to be. Except for the pathological liars, as one science correspondent called them, this seems to work a change in attitude.

A considerable number of other science specialists, however, would join a newsman who reported, "I have never been lied to, double-crossed, brushed-off, or given anything but complete co-operation and respect."

John Barbour, Associated Press correspondent who covered most of the space shots during the initial years, recalled "a certain rocket disability," which was classified during Lieut. Col. John Glenn's orbital journey but was not classified on a Ranger shot a couple of days later. Out of his wide experience at Cape Kennedy and at the Houston Manned Spacecraft Center, Barbour felt that one of the science writer's weapons when public information men exercise almost absolute power is "the ability of the individual newsman to shout, coax or embarrass them out of it on specific situations."

Control over disclosures to the news media may not only

curtail coverage but may also bottle up background vital for meaningful public discussion.

When the Soviets intimated in October, 1963, that they might eliminate themselves from the moon race, Julian Scheer, the assistant administrator for public affairs, National Aeronautics and Space Administration, sent wires to all public information officers of that agency which provided the official reaction and cautioned, "This statement by NASA on Russian statement. No further comment." As David Warren Burkett, then on the *Houston Chronicle*, said:

> No reporter could talk to officials of NASA anywhere in the world and receive anything but one stiff, formal statement.

When a contractor signs to work with NASA, the very agreement itself contains a clause restricting release of any public statements or news regarding the work without prior clearance of a NASA representative. And the contract further provides that a similar clause must be included in any subcontracts that may be negotiated.

When fire caused the deaths of three astronauts in the Apollo-I space craft in January, 1967, the news announcement was embargoed for approximately an hour and a half because NASA representatives said they could not let anything out until relatives of the dead crew men had been notified. Reporters regularly are barred from the Cape Kennedy blockhouse and thus cover tests and launchings visually or by telephone from outside.

Jim Strothman, Associated Press reporter at Cocoa Beach, Florida, at the time of the disaster, described what happened as follows:

> No reporter was present at Cape Kennedy because the National Aeronautics and Space Administration does not permit newsmen in the blockhouse during major tests

or launches. Even if we had been present, our telephones lines would have been immediately disconnected — just as telephones available to the blockhouse crew were cut off.

The wire services and local newspapers had been following the progress of the test at the Apollo I launch pad throughout the day in the only ways available to us — periodic telephone calls from a NASA information officer inside the blockhouse to the NASA press center, located 10 miles away outside Cape Kennedy gates, coupled with information through unofficial channels. Cape Kennedy is a military installation and newsmen are not permitted to roam freely.

When official NASA reports from the blockhouse suddenly ceased, conflicting rumors occasionally did slip through NASA's cloak of security.

But what reputable newsman is going to put out bulletins on a story of such magnitude when his only foundation is unattributable, unconfirmed and conflicting rumors?

While it is true that the space program is closely related to some defense activities and that the government space agency has some vested interest in not revealing every aspect of what it does, such restraints contain fertile ground for censorship and make it harder for the public to learn some of the less propagandistic facts about the expanding and expensive programs to reach the stars, all of which are financed, so far, by the taxpayers' money.

Even when censorship or outright misinformation is not present, lack of accommodations in the current public relations traditions can deter publications from covering science developments fully and at first hand. When nuclear bomb testing was conducted in the Pacific during 1962, American newspapers largely passed the story by. That June, at least one responsible newspaper, *The Washington Post*, had second thoughts about not giving its readers more information and administered an editorial chastisement that included itself:

> Looking back on the Pacific test series, we wish that all of us covering the news had sought more energetically for access to all the pictorial and written coverage that could have been permitted without endangering military security so that the American people might have had kept before them more consciously the great public issues involved in atmospheric thermonuclear testing.

Some of the more confusing aspects of bureaucracy occur when officials act as if, as one science writer described it, "they're guarding secrets too delicate for even the Atomic Energy Commission or the Department of Defense."

To illustrate what may happen in non-space fields, Stuart H. Loory, then with the *New York Herald Tribune,* said it took him two months to get a story about the National Institute of Allergy and Infectious Diseases program to develop vaccines against respiratory diseases, whereas, with "a modicum of cooperation," the article could have been done in a week.

In the fall of 1963, Loory spotted a short item in a National Institutes of Health house magazine asking for volunteers with young, virile colds who would be paid for submitting to blood tests and throat swabs. Inquiries showed it was part of the Vaccine Development Program. When Loory telephoned the scientist in charge for an appointment, the researcher suggested that he check with the public information office. Loory did, but the PIO explained he did not think he could disturb scientists at their work and offered to fill in the reporter. Although the correspondent wanted to talk to the scientist, an interview eventually was arranged with an NIAID associate director. Loory described the interview as follows in the *NASW Newsletter* (March, 1964):

> The session [which lasted an hour and a half] was filled with a reasonable amount of general information —

enough to write a once-over-lightly story but not enough
for a major feature such as I planned. The interview
finished, I asked for the tour of the laboratory.

Impossible, they said in unison. "You can't disturb the
scientists in their laboratories." Once again I appealed the
ruling to NIH headquarters but this time even that
maneuver did not work. The scientists, all working in
unclassified government laboratories and paid with government funds, stood determined to keep their doors
closed and all the pleading of public relations experts
and administrators could not coax them open.

After further appeals, Loory did tour the laboratory, but he
found the key scientists "conveniently absent" and the technicians notably reticent about discussing their work. A public
information man supplied a list of grantees working on the
Vaccine Development Program and Loory obtained enough
information to write a story by contacting Children's Hospital
in Washington, one of those institutions receiving a federal
grant.

Government agencies by no means have a monopoly on raising external barriers to communication of science news. Corporations have special reasons for wanting to delay publication of information, and science writers generally will go along
with them if the reason seems valid, provided, of course, the
news does not leak to a competitor during the interval. Nor do
many science journalists object when a drug manufacturer, for
instance, wants to withhold information of value to his competitors. But the journalist can become irate, as Walter Sullivan, science editor of *The New York Times,* explained at a
1963 University of Pennsylvania symposium:

It is only where information is sought and denied that
the science writer becomes indignant, particularly if he
suspects its concealment is because the results are unfavorable. Nor does he take kindly to the view that a
large part of the research community should evaluate a

result before it is publicized, for this is impractical and is very likely to result in leaks that get it wrong, so to speak, doing harm to the press, to the public and to science.

Our research friends must remember that reporters live in great measure for beats. They will continue to do so as long as ours is a free and open society. . . .

But the science reporter lives for more than beats. Perhaps his greatest satisfaction is to find that he has in some way enlarged the horizons of his readers, given them a deeper insight into the nature of the universe, of the atom, of life and its chemistry, and of the way science goes about exploring such matters. It's a great job, and our success in it is measured not only by our own ability, but, above all, by the help, guidance and inspiration that we get from combatants, so to speak . . .

At the same Philadelphia symposium, Arthur J. Snider of the *Chicago Daily News* talked about the science stories that turn sour — later. He explained:

My concern is that the record would show that 90% of the stories we have written about new drugs have gone down the drain as failures. We have either been deliberately led down the primrose path or have allowed ourselves through our lack of sufficient information to be led down the primrose path.

The remaining 10% of the drug stories we've written have had to be subjected to reevaluation. No drug comes out full-blown without its potential later revised. I think our mistakes arise from the fact that we are lacking knowledge on some factors in the experiment. We may lack knowledge on how the experiment was designed. We may lack knowledge on the biases that affect investigators. We may not know enough about curbing the built-in enthusiasm that an investigator may develop in his work. . . .

I wish I had more knowledge about statistics. I wish the scientists I deal with had more knowledge about statistics.

Part of these gaps in information and knowledge can be filled by public relations men — if they can resist the opportunity to plug a product, as the saying goes. The honest, reliable information officer can be inestimably valuable to the science writers but he must put his integrity (and most of them do) above the price he gets for his promotions. But even when the public relations man provides information and news with a minimum of self-advertising, his bosses may forget their roles and intervene to exert such pressures as they possess. This is unfortunate; they just mess up a job being done satisfactorily and almost always contaminate the final results.

When a booster pump was implanted in a 65-year-old man in Houston, Texas, in 1966, and surgeons called it an "artificial heart," *The New York Times* (April 22, 1966) editorially praised the "significant experiment" which offered "reason for cautious long-term optimism, not false and premature hope." *The Times*, however, commented:

> The front-page headlines and the radio bulletins that yesterday told millions about an "artificial heart" being implanted into a human being undoubtedly represented a public relations man's dream. But others will wonder about the propriety and ethics of publicizing a highly experimental procedure, with its concomitant raising of possibly unjustified hopes among thousands suffering from heart ailments.
>
> The atmosphere of sensationalism surrounding this important episode is far removed from the properly sober and cautious normal procedures of science, particularly of the sciences involved with human lives. The fact that all references to an "artificial heart" misrepresented the nature of the device involved must also add to the im-

pression that this was a classic case history of how the announcement of such matters should not be handled.

The "Madison Avenue" approach, as some might call these trends, has been adapted for publicizing the results of scientific and medical research as well as for selling products and glorifying both politicians and ideas. Some have called the exponents of this style "hucksters in the temple" but Watson Davis, long-time director of Science Service, assessed the problem as follows:

Press conferences and cocktail parties are devices adopted for the purpose of obtaining scientific publicity. Representatives of industrial concerns, particularly, seem to think that supplying science reporters with liquor or arranging pleasing junkets out of the office is an effective way of getting publicity.

Perhaps it is, but there has been a reaction against succumbing to such blandishments. The cocktail party or the junket can be a form of bribery, although the public relations boys will deny this vehemently. Obviously, if any science reporter is going to sell out to public relations, the price will have to be considerably higher than a few dollars for cocktails, dinners, and airplane fares. Some newspapers have covered these PR events but they have insisted upon paying the way of their representatives; others have felt that their reporters are strong enough in ethics to withstand any temptation and bias.

13

TRAINING FOR SCIENCE WRITERS

Training to become a successful science journalist with the mass media rests on the two giant columns that also support such reporting itself: (1) Sufficient background in science to ask intelligent questions and to comprehend the answers and (2) high competence in writing the information received so that it will appeal to the non-scientist public.

Little disagreement exists about these two stark prerequisites, but there are all sorts of disputes when it comes to implementing them. As recently as 1964, a former president of the National Association of Science Writers who had won numerous prizes for his reporting excellence on a metropolitan newspaper could write, "No one, at this point, knows what background is required to make a good science writer. This is one of the unknowns that should be ascertained." And while there has been considerable tentative probing, there has been no consensus.

Some of the confusion arises because the initial pioneers, especially among newspapers and press associations, got into their assignments almost by accident as science reporting jobs developed while they already had other news assignments. Only recently have potential journalists been able to point toward a career in science writing while they were undergraduates; such a curriculum is still fairly experimental training. Today only a few beginning science writers have prepared by enrolling in such college studies as have been established. Nor should these be required for employment.

Some journalism professors favor specific science writing courses at the undergraduate level; others opt for graduate work; still others oppose pinpoint specialization at any level. If a fifth-year graduate program in science journalism is easily available and if funds to pay for such an education are readily at hand, it makes little difference where such technical and semi-professional work is scheduled. If an undergraduate himself can fuse his science knowledge and his journalism instruction, either in college or early on the job, all will be well. But if it is possible to help an undergraduate who otherwise might have serious difficulties in such adjustment, that too should be considered an educational plus, provided it does not deprive him of the more essential training in science and general journalism.

Among some of the complicated questions in training science writers are the following:

1. If and when a choice has to be made, is science background more important and to be preferred to writing abilities?

2. How should a neophyte train for an eventual job in science writing?

3. What does one do with the professional newsman who, midway in his career, wants to shift into covering the spec-

tacular space voyages and "breakthrough" developments on the frontiers of basic research and development?

4. Should science journalists be trained to become specialists with few, if any, assignments in other news fields?

Prof. James Stokley, Michigan State University journalism teacher, former staff member of Science Service, and a one-time public information man who wrote science releases, said this about the dual requirements for handling science news:

> While the science writer fundamentally is a journalist and not a scientist, he must have many aspects of the scientist. He should have sufficient scientific and technical knowledge to understand moderately technical articles and to talk with scientists concerning their work. He must understand how scientists work and be sympathetic with them and their aims. This is necessary to gain their confidence and respect, which the science writer certainly needs.
>
> Much of this he will get on the job with experience, for the best way to learn about scientists is to work with them. But he also needs general newspaper experience, which can best be obtained by working as a reporter. Some of this might be obtained by working on a good college daily. . . .
>
> Some have maintained that science training is unnecessary, because if a scientist clearly explains a technical subject to him, he can pass it on to his readers. Unfortunately, however, many good scientists are not very articulate and the science writer needs to know something about the subject to get the information from them.
>
> With the rapidly increasing complexity of science and technology, it is becoming more and more difficult for even a good journalist to acquire, solely by his own study, the broad knowledge that a science writer should have. I feel that even at the start of his career the science writer should have some basic knowledge in several scientific

areas, and this can best be obtained in college courses. While no courses now exist for many of the important things that he will be writing about in 1975, his basic knowledge should make it somewhat easier to learn about such new developments.

Some of the professional in-fighting about priorities for science versus journalism backgrounds arises over what should be the basic orientation of an individual candidate. He could concentrate on science training and hope to add writing as a sort of frosting or he could do just the opposite. Actually, since both are important, it should be reasonably simple now to do both — and omit some of the near-endless discussions. And this dual training is what universities are offering to students who seek to prepare for science writing jobs.

Undoubtedly there is some value — as Watson Davis, then director of Science Service, suggested in his 1960 American Chemical Society News Service dinner speech in connection with receipt of that year's James T. Grady medal — to having a prospective science newsman really "get his hands dirty and his mind disturbed in some sort of research laboratory." But this should not be done at the expense of developing his writing capabilities. (With anything like intelligent direction, such curtailments need not take place in training.)

At least one veteran newspaper science reporter has questioned the need for extensive classroom science preparation for becoming a science writer. Harry Pease of *The Milwaukee* (Wisconsin) *Journal* said he found no high correlation between science writing competence and considerable science training in the classroom. He added:

Certainly a science writer should have enough laboratory courses in his background to give him a feeling for the conditions and criteria which surround the scientist's work. But I submit that detailed training in specific disciplines has very doubtful value. . . .

Whatever value scientific training has for the writer

boils down then to the nebulous quality called background, which is presumed to equip the reporter to ask
an intelligent question at the right time.

I guess what I really resent is the inference that none
of us has learned anything since he got out of college.
Actually those years are a small part of a 40-year-old
man's total experience. If we are any good we read a lot
— and the reading is usually heavier than that pursued
by a college freshman taking general chem. We interview
the best men in science, and I suggest that a two-hour
talk with a man on the forefront of DNA research is considerably more educational than a semester spent listening to the lectures of a graduate assistant in zoology.

It seems to me that we are nearing the time when we
can claim professional qualifications peculiar to our trade,
rather than seeking to be regarded as scientists with an
incidental knowledge of the English language as she is
spoke. We represent the inquisitive layman, not the literate research worker. If we recognize it, emphasizing our
capacity to approach qualified sources, we'll gain respect.
If we try to pose as experts we can only appear ridiculous, in my opinion.

Making much the same point as Pease, Charles A. Scarlott,
manager of the Publications Department, Stanford Research
Institute, commented after a 1961 Conference on the Role of
Schools of Journalism in the Professional Training of Science
Writers:

An understanding of the scientific method and scientific
principles is necessary in science reporting for at least
two reasons: a) To comprehend what is to be interpreted
and to be able to set the development in the proper perspective in the total. b) To win the confidence and respect of the scientist. Without it the scientist — consciously or unconsciously — becomes uncommunicative
or worse.

A science reporter needs to understand scientific principles and methods, not be knowledgeable in the specifics of a given development. . . .

To be a skillful interpreter of science, one does not have to be experienced in all fields of science. There is a great deal of carryover between fields. But the carryover is not scientifically universal. For example, an engineering degree does not help much in reporting medical accomplishments.

In recent years, the Council for the Advancement of Science Writing, Inc., has undertaken to place a few established science journalists in laboratory positions for a month or two during their summer vacations. This has been successful enough among the experienced reporters, who know where they want to fill in their background knowledge, to establish it as a possible on-the-job approach for upgrading writers who already have long-term media jobs.

When the chips are down, editors, publishers, and broadcast executives do not often have to pick from individuals who have only either/or backgrounds in science and journalism. Few individuals seek positions without at least a little writing experience or without at least their own private science indoctrination through free-time reading. To date, when news executives have had to face this choice, they have far more frequently selected individuals with news experience, hoping that the writers would acquire rapidly the needed science backgrounding on the job.

What will happen in the future may be quite different; at least that is the opinion of Lee Hills, executive editor of the Knight newspapers. In the 1960 Don R. Mellett lecture at the University of Oklahoma, Hills said:

> I venture to predict that before many years pass our major newspapers will be able to find and willing to pay bright young medical graduates who will write about medicine, educators who will quit the campus to write about education, physicists who will desert the laboratory for the city room. . . .

If this comes to pass (as it has already to a limited degree) it may transform tomorrow's newspaper. Despite the frills and frivolities of modern America, never before have people so hankered for the fact, spun out plainly and at length.

These specialists, of course, must know or learn how to communicate with the written word. They must report as well and write as well as other trained newsmen.

Indeed, we need to tell the story of our age in simple, living language with precise meanings. We will not only inform but we will also educate a generation. And as we develop greater skills for reporting and interpreting these events factually and accurately, journalism will at last earn the status of a genuine profession.

Whether this prediction will come true depends on finding and training scientists who "report as well and write as well as other trained newsmen." Most contemporary science journalists believe such people will be hard to locate. Media work demands too much background and talent to be filled by academic dropouts whose poor grades in technical courses prompted them to seek other vocations. To date, a few have switched into communications from scientific, technical, or medical schooling because they revolted against the behavior patterns expected from people in these professions. If and when scientists and physicians with the dual qualifications outlined by Hills are discovered in the future, certainly well-paying jobs will be awaiting them if they really want to forsake laboratories for the newsrooms.

As reported in Chapter 6, science writers now on the job tended to concentrate on mathematics and the physical sciences during their own college years, but this may reflect only a historical fact and serve as a poor guide for the future.

But if they do not aim at becoming experts in any science, then what courses should the prospective science writers study? Or, for that matter, which topics should they include in their home readings?

When general science courses and books are truly rigorous

in content and conception and are far way from the "Mickey Mouse" tradition of easy credits, they can supply some needed background. But the newsman should avoid over-simplified surveys, whether courses or books. If the science writer attains a broad background, then he will be able to handle assignments across many fields.

If the student (or veteran shifting his goals while on the job) will spend enough time to reach at least one frontier in science, he will be better able to experience the agony and ecstasy of even minor "breakthroughs" achieved by the researchers he interviews. But it is not required. If he concentrates his science training in a single field in which he has a special interest, he is bound in, when he leaves his specialty, to the same degree as any general assignment reporter who covers a political conference one day, a science meeting the next, and a teachers' convention the third.

Readings or courses in the history and philosophy of science are considered "essential for adequate training" by such people as Dr. Chauncey D. Leake, former head of the American Association for the Advancement of Science and long interested in popularizing scientific and medical achievements.

At a 1961 Conference on the Role of Schools of Journalism in the Professional Training of Science Writers, held by Science Service in Washington under a National Science Foundation grant, representatives from more than a score of journalism schools and departments went on record as opposing "the formulation of arbitrary curricula in education for science writing"; most participants felt programs should depend upon an individual's special talents and aims.

While specialization is taking place on some of the larger metropolitan daily staffs, such jobs are rewards of experience and do not often go to those who are not among the professional elite.

For the experienced reporter who wants to specialize in science at mid-career, the question of on-the-job training is complicated.

In an article in *The Quill* (August, 1960), Blair Justice,

Texas science writer now on *The Houston Post,* offered this advice for a reporter who wanted to shift into science writing:

> The most obvious way for a reporter to gain a grasp of science, of scientific language and concepts is to go back to school. . . . I myself have been back at school for four years, taking one course a semester in an evening college and concentrating on just one field I think is applicable to my job in writing medical news — psychology.

But Justice realistically pointed out that most part-time science specialists might not want to undertake such rigorous postgraduate training.

Short courses and seminars for established science writers became fairly popular during the 1960's. This was not a new idea, but simply an expansion. For a number of years, going back to 1953, the American Cancer Society and later such other fund-raising organizations as the National Foundation and the National Tuberculosis Association had conducted tours and conferences to publicize what they were doing to overcome diseases. The ACS annual "tour" (now held at a single site rather than at a different location daily) became a "must" assignment for some science reporters on metropolitan dailies and always provided considerable newsworthy publicity just ahead of fund-raising campaigns.

Among government agencies, the National Aeronautics and Space Administration and the National Institutes of Health of the U. S. Public Health Service have sponsored briefings for science and medical writers at their installations. In advance of manned satellite launchings, NASA not only has held news conferences but also has supplied packets of mimeographed background materials.

Since 1960, the National Science Foundation has financed a wide range of briefings, which varied from a single day to up to two weeks. These sessions were to provide either background on basic sciences or information on recent developments and they often were held on or near college campuses.

Thanks to NSF financial assistance, the American Institute of Physics has held more than a dozen one-day briefings on various aspects of physics and then spread the effectiveness of its programs by distributing glossary booklets on each subject discussed. Some NSF conferences were for full-time professionals but often they were for "young veterans" or even beginners. At least one was held for public information men, especially from colleges in the Southeastern United States, on the announced assumption that many of the smaller papers in that region depended heavily on press releases for their science news content; thus, to improve the quality of these releases was to bring better coverage in print. After half a decade of operations, at least several thousand "billets" have been provided at these NSF-financed meetings. Just how many individual reporters actually attended was difficult to estimate since some newsmen participated in several different sessions.

Since 1963, CASW has held an annual week-long conference on "New Horizons in Science" to up-date many of the elite professional science writers who already had acquired much essential background. For general assignment reporters who might have to handle medical emergencies, one pharmaceutical concern financed one- and two-day seminars on the theory that sharing general information more widely could not help but improve newspaper coverage when no specialist was around to handle the story.

Although some summer writing conferences have been scheduled by colleges and universities, these turned out to be patronized more by technical writers, especially those from industry, rather than by science journalists with the mass media. Science writers, especially those with considerable newsroom experience, have been allergic to sessions concerned with how-to-do techniques, preferring rather to learn about science, technology, and medicine when they got together.

Under a series of grants from the Carnegie Corporation of New York, CASW also has pushed an on-the-job training program that is a miniature course itself. The plan was not designed to produce full-time science correspondents (although it has in some cases) but rather to give participants the factual background for handling science news as it comes along among varied assignments, to engender critical and self-confident judgment in the field, and to give them a knowledge of some solutions to the special problems of science writing. Within less than five years, CASW has provided such training for several hundred newsmen and has seen some of them move increasingly into more intensive coverage in their own communities.

Among the features of this CASW on-the-job program are:

1. Supplying each participant with selected reference books, some for him and some for the newspaper's library.

2. Providing annual subscriptions to magazines and journals that will give the newsmen a sense of participation in science coverage and will keep them abreast of current advances.

3. Making available a series of readings on both science and science journalism, including quarterly selections of representative science news articles that have appeared in papers across the country. Some of those reprinted had won national awards.

4. Assigning an established science writer to work as adviser and counselor. This professional not only went directly to the beginner's office to help him but also spent time with him later when both attended regional science or medical meetings.

All of this indicates that during the past decade, considerable time and money have been expended to up-grade the basic and technical backgrounds of those reporters on the job. If one can accept the admittedly biased opinions of those connected with these various projects, they generally have succeeded in raising the level of scientific literacy of those who translate much of the world's news for the general public.

A 1966 survey of schools and departments of journalism by Henry A. Goodman, CASW executive secretary, showed that 26 of the 127 responding institutions either had specific courses or arrangements for study of science writing or background work in the sciences. An additional 77 schools and departments reported that they included science writing as part of other instruction.

Among the going-back-to-college graduate programs that have attracted wide and generally favorable comments are those at the University of Wisconsin and at Columbia University. In addition, the Nieman Foundation for journalists at Harvard University has included science writers among those newsmen invited to spend a year browsing at Cambridge in the academic fields of their greatest interests; during the early years one science reporter was selected almost annually and more recently this has been insured by a special fellowship in science writing underwritten by Arthur D. Little, Inc.

Since 1948, the University of Wisconsin School of Journalism and the University News Service jointly have offered at least two graduate fellowships yearly for training in science journalism. The News Service assigns the fellows to work with full-time public information men covering science, engineering, and medicine, and thus the students supplement their course work with on-the-job experience. Many of their reports, after editing and possibly rewriting, are distributed through public relations channels, sometimes with the graduate students' bylines. Fellowships are granted for either 10 or 12 months and carry a monthly stipend.

The Madison campus also was selected to launch a training project in reporting social sciences; it started in the fall of 1964 and was financed by grants from the Russell Sage Foundation. Course work was designed to serve two purposes: to give social science background to experienced newsmen and to provide writing opportunities for already-trained social scientists.

Although the initial Russell Sage Fellows sampled courses from many departments, they tended to concentrate in political science, sociology, and economics. All the Fellows took work in methods of social science research or in statistics; all attended a series of seminars in which various social science fields were explored with experts and the newspaper potentialities discussed; and each Fellow was expected to undertake a research-writing project. Among those for the first year were "Miscegenation and the Court," "The Bias in our Crime Statistics," and "Defending the Guilty — Plea Bargaining and the Need for Counsel."

After completing their year of course work, seminars, and special trips to such professional meetings as the annual conventions of the American Psychological Association and the American Sociological Association, the Fellows returned to their newspapers. Among the five dailies represented by the initial class members were the *St. Louis Post-Dispatch* and *The Wall Street Journal*. Most of the participants wrote special series on topics in the social sciences during the early months back on their papers, but one former Fellow reported that he did his book research at home where his editor would not see his time "wasted" on work the boss felt was unneeded.

Dr. Charles E. Higbie, coordinator of the Russell Sage program, said the initial years had produced fruitful results on the part of both newsmen and social scientists. Then he added:

> I would say that the need for the specialized program has been confirmed, but in more realistic terms. In addition, immediate dividends in the form of more perceptive writing have already accrued. However, more basic and more important results in many forms will indicate the value of the program over a longer period of years.

At Columbia University's Graduate School of Journalism since 1958, an advanced science writing program has undertaken "to help raise the quality and to increase the quantity of science writing as a means of broadening public understanding of science." The program has provided an opportunity for up to 10 journalists or, in some cases, scientists, to spend an academic year broadening their knowledge of science and sharpening their reporting and writing techniques. Participants have usually been college graduates with good academic records and at least three years of professional experience in mass media reporting or in scientific research. Most of the Columbia Fellows have been in their early thirties and the experience level has been well over five years. Students have sought to fill gaps in their general scientific backgrounds left during their under-graduate days and, in some cases, to provide advanced work in specialized fields.

The Columbia project was liberally supported in excess of half a million dollars, largely by grants from the Alfred P. Sloan Foundation and the Rockefeller Foundation. Fellows receive full tuition and fees plus stipends of up to $4,400 for travel and living costs. Stipends are based on the individual's salary, family, etc.

Of the 70 Fellows accepted up through the 1966–67 academic year, 37 came from newspapers, five from wire services and syndicates, seven from magazines, four from radio and television, and four were free-lancers. As a counterpoint variation, five had science backgrounds and broadened their writing abilities during their year on Morningside Heights. Of the eight others, two came from foreign governments, one from a United Nations agency, one from advertising, one from university administration, one from university public relations, and two from book publishing. The largest shifts in employment that came as a result of the program were a decline to 20 now with newspapers and an increase to 18 now on magazines.

Assessing the Columbia program, Prof. John Foster, director, explained:

> One of the most important reasons for the success of the program has been the unusually fine cooperation of the university's administration and scientific faculties. Because of their interest, it has been possible for the Fellows to study at the Medical Center, the Psychiatric Institute, the School of Public Health, the Graduate Faculties, the School of Engineering and Applied Sciences, specialized units such as the Lamont Geological Laboratory, the Electronics Research Laboratories, and the Watson Scientific Computing Laboratory — almost anywhere and everywhere there are courses or seminars of value to mature men and women who want to learn.
>
> An important part of the program has been the field

seminars in which the Fellows make one to five-day visits to leading government, industrial and other research and development centers (such as Brookhaven National Laboratory, du Pont, Bell Telephone Labs, IBM's Watson Research Center, General Electric Space Lab, Cape Kennedy and Wright-Patterson Air Development Center). There has been ample time to discuss in detail with the scientists who are doing the work, college research in a wide variety of fields.

In 1966, the Alfred P. Sloan Foundation gave a founding grant of $1,000,000 to establish the Institute for Studies of Science in Human Affairs at Columbia. A quarter of this fund was earmarked for continuing support of the Advanced Science Writing Program in the Graduate School of Journalism and for that program's progressive integration with the Institute's work.

In a much more restricted area, the Syracuse University School of Journalism has launched a three-year pilot training program to prepare information specialists for state departments of mental hygiene and public relations workers at institutions and community psychiatric centers. Funded by the National Institute of Mental Health, a grant in 1964 provided four fellowships each year with a stipend of $3,000 plus full tuition.

Noteworthy is the fact that these graduate level training activities have been supported by grants from foundations and government. What may happen when such financial aid is discontinued is a question that only the future will answer. That "moment of truth" should provide some measure of how essential the administrators of mass media and the general public believe it is to have well and better trained science journalists from the college campus.

14

WHAT OF THE FUTURE?

A glib, happy prediction for a steady onward and upward trend in science journalism during the years ahead would be pleasantly euphoric but it would lack realism.

Certainly science reporting can — and hopefully will — become better in both content and in style of presentation.

Undoubtedly it will receive more news space in newspapers and magazines and more time on radio and television schedules, especially when men first land on the moon, surgeons increasingly replace defective human organs, and the double helix of the "code of life" with its potentialities for improving inheritance, finally is charted in full.

Probably journalists will pay more attention not only to research and development but also to science's role in the national economy and in government, thus creating news assignments of eco-science and politico-science coverage.

Possibly science commentators and columnists — real interpreters of contemporary science — will approach the status of present-day political writers.

But, unless the mass media undergo a complete change in philosophy — of which there is little present indication — science writing will not replace the schools as the primary educational channels for the nation's young; weekly science pages and possibly the new columnists will supplement classroom instruction for both youngsters and adults as they supply more informed backgrounding for current events. Yet, even during the next generation, it seems unlikely that science coverage will become so popular as to surpass sports sections or business and financial stories as major news departments in daily papers, news magazines, radio, and television.

A few newsmen have questioned whether science reporting is not already approaching a plateau, at least in numbers employed on daily papers. They admit that room for expansion exists on magazine staffs and in electronic journalism but they have doubts about their own print field. These correspondents seem unduly pessimistic but the explosion which saw the number of professional science writers increase by approximately ten times in the past third of a century is unlikely to be repeated in the years remaining before the twenty-first century.

If the American people are to get "better" science news coverage in the future, just what does it mean?

It means that there will be more science news in sheer volume in print and on the air. Practically every study during the past several decades has shown that increasing amounts of such news are being delivered to mass audiences; this is true for all the major media channels, including paperback books on current topics. As the space program moves into even more suspenseful events, as research goes into new and exciting areas, and as development brings more gaudy and handy products, the media will follow. Because there will be longer gaps of "free time" as the space voyages lengthen, for instance, more explanations and background will inevitably go on the air and into print.

This leads to a second trend: more analyses, interpretive articles, and Sunday or weekly comments. In addition to filling time between visual space happenings, the public's need for this information has been established and many science journalists and some editors and producers are trying to do something to change and to improve performances. No longer will it be likely that a television commentator faced with an open microphone and no immediate event to report will blurt out, as one did in 1965, "Oh, it all gets so complicated." And this will take place because responsible professionals do not like to expose such ignorance and so will furbish their own knowledge and eventually pass some of it on to the public.

Extension of the time involved in space flights will eliminate the early traditions of continuous television coverage from lift-off to splashdown. But even this revision in programming will leave plenty of time for meaningful background and interpretation. Newspapers and news magazines may be expected to pursue their reporting-in-depth coverage — and possibly increase it for what their staffs consider super-colossal spectaculars. But there is more to science than shooting for the moon and the stars, and here backgrounding is badly needed if the spot news is to make much sense for the general readers and viewers.

Victor Cohn, science reporter for the *Minneapolis Morning Tribune* and highly regarded both by scientists and by his peers, wrote in *Science* (May 7, 1965):

> We science writers, except for an exceptional few, fail to pay enough attention to basic research, and we too often fire out news of new discoveries, or what we call discoveries, without connecting them with the main body of knowledge and the basic work that has gone before.
> We over-use a bagful of clichés, like "major break-

through" and "giant step forward." . . . We especially over-enthuse on medical "discoveries." . . .

For all these reasons we are not truly *covering* science and technology and their huge, terrifying, and inspiring impact. We are missing too many of the big stories of our time through daily preoccupation with trivia. We are ignoring the social and behavioral sciences almost completely.

Professional science writers like Cohn know what is wrong and want to do something to correct the errors. With help from their bosses and from the scientists who are their news sources, real progress could be achieved reasonably soon.

Science coverage in the future also should be effected by the general reevaluation of journalistic traditions that is taking place through efforts to provide more readable and listenable reporting. Writers, editors, publishers, and producers talk about the "straitjackets" of some common practices and are seeking to struggle out of many of them. For instance, can science not be written in terms of human interest without diluting the content to superficiality? Cannot more time be devoted to documentaries instead of dismissing a news development with 90 seconds during a newscast?

More attention will be given to the philosophical concepts of basic research that underpin the spot news events and to those tangential areas where science meets economics and politics. Already some metropolitan dailies are printing stories on how much money is going to research and development for a specific industry and predicting what the future potential profits could be. A few magazines — especially *Science,* the weekly publication of the American Association for the Advancement of Science — regularly follow to where science and government meet, not infrequently in minor combat for appropriations and allocations of grants for research and multi-million dollar installations. Whether politicians or scientists admit it, politico-science coverage will tell the story of a new potential in allocating funds which could, unless kept under constant scrutiny, surpass the traditional rivers and harbors appropriations as "pork barrel."

Other new areas of science journalism are ready for coverage and hints are at hand as to where reporters will go to find future stories. As mentioned in an earlier chapter, the social sciences seem ripe for more extensive coverage and this should include not only the more traditional ones such as economics, political science, and geography but also the newer ones such as anthropology, psychology, and sociology. But these changes will not come without some missionary work in the media news rooms and some recognition, probably agonizing, by the social scientists that they lack, as Dr. Ernest Nagel, John Dewey professor of philosophy at Columbia University, explained, "the almost complete unanimity commonly found among competent workers in the natural sciences as to what are matters of established fact, what are the reasonably satisfactory explanations (if any) for the assumed facts, and what are some of the valid procedures in sound inquiry."

Race relations with their occasional riots and violence and the Johnson Administration's Great Society with its educational innovations and "war on poverty" have zoomed on to front pages across the country. These news developments and the media's efforts to make them more understandable to the public that reads and listens undoubtedly will bring increasing coverage of the social scientists and their research findings, just as the depression stimulated extensive reporting of the government's entry into regulation and controls of business and society during the 1930's. Urban renewal projects merit more than just stories about tearing down buildings and then replacing them as if they were simply news for real estate pages. Educational changes include far more than just adoption of annual budgets, meetings of the Parent-Teacher Association, and hiring of new teachers for the elementary and secondary schools.

As "Big Science" grows, it impinges on many other areas and this generates reportable developments of probable interest to large segments of the non-scientist public. If science has become a major force in contemporary life, as argued in Chapter 1, then intelligent laymen will want to know more about it, even if the motivation is sometimes other than a quest for knowledge or an effort to share in an exciting and stimulating pursuit of excellence.

Impressive as are the reasons for better preparing the public to participate in the decision-making process, we must keep in mind that science, even at its peak efficiency, will not give all the answers. Obviously it is important that those who have a hand in making choices know what the alternatives are. But we must never forget that many decisions have their subjective elements and that science can be a weak guide in these areas.

One compelling reason for anticipating improvement in the quality, if not the quantity, of science reporting by the mass media in the near future has nothing to do with increased dedication of journalists, desires of scientists, or civics of decision-making. It relates to the new audience that has been created recently. Due to the new science and mathematics curricula that were introduced in the late 1950's and early 1960's, a whole generation of former high school students, better trained than any previous one in the nation's history, is moving toward adulthood and shortly will become the new parents. Never have so many Americans been introduced to science, technology, and medicine during their teens. The "bright 17-year-old" of 1960 who completed four years of the new mathematics and at least three years of the new science courses while in high school will become the 27-year-old adult of 1970.

Educational surveys show that for the late 1950's and early 1960's, approximately three-quarters of the rising tide of high school students took biology, slightly more than two-thirds studied general science, somewhat more than one-third enrolled in chemistry, and slightly less than one-quarter had physics. This meant that for 1962, for instance, close to 2,500,000 high school students took biology courses during that single year. Thus one could project that approximately 25,-000,000 or more of the young adults during the 1970's would have studied that most popular of all high school sciences, biology, during the decade of the 1960's. This was approximately three times the number two decades earlier. When they become adults in the 1970's, this number of biology-trained, former high school students of the 1960's will come close to one out of every five persons in the entire grown-up population.

The improvement in secondary school science and mathematics teaching also has had explosive impact outside the country's high schools and it has not been confined just to these subject areas. For instance, Dr. James R. Killian, Jr., chairman of the Massachusetts Institute of Technology Corporation and first science advisor to President Dwight D. Eisenhower, pointed out in 1965 that the new science and mathematics curriculum reforms had spread to other fields:

> This revolutionary improvement in American education is not only invigorating and upgrading the quality of teaching in pre-college schools; it is now enabling, and indeed, forcing the colleges to advance the level of their studies and to give really creative attention to the reformation of their curricula. . . . Allied to this curriculum reform are the brilliant new college texts and other teaching materials which are the work of leading scientists resident in institutions deeply engaged in research.

What the better informed young adults will bring to their newspaper and magazine reading and their radio and television listening could force a communications revolution such as has seldom been equaled. The finding that science courses increase science news consumption from the mass media is established. The glib over-simplifications and heavy sensationalism that satisfied the young adults' parents will bring abusive letters to editors and, unless corrected, a contempt for the popular media that could launch catastrophic changes in how people get their news, not only about science but about all types of events.

Most of the United States already has a multi-media communications system with little reliance on any single medium except in a few scattered, isolated communities. Thus, interest on the part of one medium often is reflected in the performances of the others. With science and technology staging more and more dramatic activities, television will have the centuries-old news value of conflict (or "Man Against Nature") in space spectaculars to justify increased coverage. When television adopts this pattern, print channels will have to reassess their own operations and compete on the level where newspapers and magazines can win most points in competition with electronics: interpretation and explanation of behind-the-scenes, non-visual backgrounding. This does not mean that printed media will omit spot news coverage but simply that they will emphasize the background as well as the foreground of current events. And with this should come coverage of the less dramatic but probably more important advances in research.

With a little bit of luck, the next decade or so could see the development of science columnists with each having a strong possibility of his own following just as political commentators now have their enthusiastic claques. But before that happens, newspapers will have to recruit candidates for the assignment who have both a stronger foundation in science and a greater fluency in writing than is found in many contemporary science journalists or, for that matter, in most scientists. Occasional columns of comments would be relatively easy for a number of professionals, but to do a three-a-week performance of worthwhile evaluations and interpretations would demand an overwhelming versatility.

Now that science journalism, even for the electronic media, is not brand new, it may be expected that more and more of those preparing to write about science, technology, and medicine will start in college, either as undergraduates or as graduate students. While some who become eventual professionals will transfer after years in the newsroom, led to the field by only their own personal inclinations, the normal pattern in the future probably will not be the stagger-in process which all the initial pioneers followed when they became science writers. Training for science journalism has become too important for a hit-and-miss educational approach. The intelligently trained individual always will have a chance to prove himself in science writing for the mass media, just as he does in other journalistic areas. But this kind of news is too important to allow any system that brings many "misses."

About the Author

Chairman of the Department of Journalism at New York University (1957–1963), and currently Professor of Journalism at N.Y.U., Hillier Krieghbaum is a former newspaperman and a specialist in popularizing science information. Before joining the N.Y.U. journalism faculty in 1948, he taught at the University of Oregon (1946–1947) and at Kansas State College (1938–1942).

A native of South Bend, Indiana, Professor Krieghbaum was graduated from the University of Wisconsin in 1926 with a bachelor of arts degree in journalism. He received his master of science degree from the Medill School of Journalism at Northwestern University in 1939.

In 1942, Professor Krieghbaum was head of the UP Washington bureau news staff covering civilian war agencies, and during the postwar reorganization of the Veterans' Administration, he was a public information specialist for the department of medicine and surgery.

Professor Krieghbaum is the author of five books on journalism: *American Newspaper Reporting of Science News* (1941), Kansas State College; *Facts in Perspective: Editorial Page and News Interpretation* (1956), Prentice-Hall; *When Doctors Meet Reporters* (1957), New York University Press; *Science, the News, and the Public* (1958), New York University Press; and, as coauthor with Edmund C. Arnold, *The Student Journalist* (1963), New York University Press.